Introduction to Elementary Computational Modeling

Essential Concepts, Principles, and Problem Solving

Chapman & Hall/CRC
Computational Science Series

SERIES EDITOR

Horst Simon
Deputy Director
Lawrence Berkeley National Laboratory
Berkeley, California, U.S.A.

AIMS AND SCOPE

This series aims to capture new developments and applications in the field of computational science through the publication of a broad range of textbooks, reference works, and handbooks. Books in this series will provide introductory as well as advanced material on mathematical, statistical, and computational methods and techniques, and will present researchers with the latest theories and experimentation. The scope of the series includes, but is not limited to, titles in the areas of scientific computing, parallel and distributed computing, high performance computing, grid computing, cluster computing, heterogeneous computing, quantum computing, and their applications in scientific disciplines such as astrophysics, aeronautics, biology, chemistry, climate modeling, combustion, cosmology, earthquake prediction, imaging, materials, neuroscience, oil exploration, and weather forecasting.

PUBLISHED TITLES

PETASCALE COMPUTING: ALGORITHMS AND APPLICATIONS
Edited by David A. Bader

PROCESS ALGEBRA FOR PARALLEL AND DISTRIBUTED PROCESSING
Edited by Michael Alexander and William Gardner

GRID COMPUTING: TECHNIQUES AND APPLICATIONS
Barry Wilkinson

INTRODUCTION TO CONCURRENCY IN PROGRAMMING LANGUAGES
Matthew J. Sottile, Timothy G. Mattson, and Craig E Rasmussen

INTRODUCTION TO SCHEDULING
Yves Robert and Frédéric Vivien

SCIENTIFIC DATA MANAGEMENT: CHALLENGES, TECHNOLOGY, AND DEPLOYMENT
Edited by Arie Shoshani and Doron Rotem

INTRODUCTION TO THE SIMULATION OF DYNAMICS USING SIMULINK®
Michael A. Gray

INTRODUCTION TO HIGH PERFORMANCE COMPUTING FOR SCIENTISTS
AND ENGINEERS, Georg Hager and Gerhard Wellein

PERFORMANCE TUNING OF SCIENTIFIC APPLICATIONS, Edited by David Bailey,
Robert Lucas, and Samuel Williams

HIGH PERFORMANCE COMPUTING: PROGRAMMING AND APPLICATIONS
John Levesque with Gene Wagenbreth

PEER-TO-PEER COMPUTING: APPLICATIONS, ARCHITECTURE, PROTOCOLS, AND CHALLENGES
Yu-Kwong Ricky Kwok

FUNDAMENTALS OF MULTICORE SOFTWARE DEVELOPMENT
Victor Pankratius, Ali-Reza Adl-Tabatabai, and Walter Tichy

INTRODUCTION TO ELEMENTARY COMPUTATIONAL MODELING: ESSENTIAL CONCEPTS,
PRINCIPLES, AND PROBLEM SOLVING
José M. Garrido

Introduction to Elementary Computational Modeling

Essential Concepts, Principles, and Problem Solving

José M. Garrido

Kennesaw State University
Georgia, USA

CRC Press
Taylor & Francis Group
Boca Raton London New York

CRC Press is an imprint of the
Taylor & Francis Group, an **informa** business

A CHAPMAN & HALL BOOK

MATLAB® is a trademark of The MathWorks, Inc. and is used with permission. The MathWorks does not warrant the accuracy of the text or exercises in this book. This book's use or discussion of MATLAB® software or related products does not constitute endorsement or sponsorship by The MathWorks of a particular pedagogical approach or particular use of the MATLAB® software.

CRC Press
Taylor & Francis Group
6000 Broken Sound Parkway NW, Suite 300
Boca Raton, FL 33487-2742

© 2012 by Taylor & Francis Group, LLC
CRC Press is an imprint of Taylor & Francis Group, an Informa business

No claim to original U.S. Government works

Printed in the United States of America on acid-free paper
Version Date: 2011915

International Standard Book Number: 978-1-4398-6739-6 (Paperback)

This book contains information obtained from authentic and highly regarded sources. Reasonable efforts have been made to publish reliable data and information, but the author and publisher cannot assume responsibility for the validity of all materials or the consequences of their use. The authors and publishers have attempted to trace the copyright holders of all material reproduced in this publication and apologize to copyright holders if permission to publish in this form has not been obtained. If any copyright material has not been acknowledged please write and let us know so we may rectify in any future reprint.

Except as permitted under U.S. Copyright Law, no part of this book may be reprinted, reproduced, transmitted, or utilized in any form by any electronic, mechanical, or other means, now known or hereafter invented, including photocopying, microfilming, and recording, or in any information storage or retrieval system, without written permission from the publishers.

For permission to photocopy or use material electronically from this work, please access www.copyright.com (http://www.copyright.com/) or contact the Copyright Clearance Center, Inc. (CCC), 222 Rosewood Drive, Danvers, MA 01923, 978-750-8400. CCC is a not-for-profit organization that provides licenses and registration for a variety of users. For organizations that have been granted a photocopy license by the CCC, a separate system of payment has been arranged.

Trademark Notice: Product or corporate names may be trademarks or registered trademarks, and are used only for identification and explanation without intent to infringe.

Visit the Taylor & Francis Web site at
http://www.taylorandfrancis.com

and the CRC Press Web site at
http://www.crcpress.com

Contents

List of Figures		xv
List of Tables		xix
Preface		xxi
About the Author		xxv

I Understanding Problem Solving — 1

1 Understanding Word Problems — 3
- 1.1 Introduction ... 3
- 1.2 Nouns and Verbs ... 4
- 1.3 Problem Decomposition ... 6
- 1.4 The Language of Computational Problems ... 8
  - 1.4.1 Terms for Computed Values ... 8
  - 1.4.2 Implied Phrases ... 9
  - 1.4.3 Units of Measure ... 9
  - 1.4.4 Conditions ... 11
  - 1.4.5 Repetition ... 13
- 1.5 Objects ... 14
- 1.6 Problems with Many Numbers ... 15
  - 1.6.1 Working with Patterns ... 16
- 1.7 Summary ... 17
- Key Terms ... 17
- Exercises ... 18

2 Problem Solving and Computing — 21
- 2.1 Introduction ... 21
- 2.2 Programs ... 21
- 2.3 Data Definitions ... 22
  - 2.3.1 Name of Data Items ... 22
  - 2.3.2 Data Types ... 23
  - 2.3.3 Data Declarations ... 24
- 2.4 Programming Languages ... 24

	2.4.1	High-Level Programming Languages	24
	2.4.2	Interpreters	25
	2.4.3	Compilers	25
	2.4.4	Compiling and Execution of Java Programs	25
	2.4.5	Compiling and Executing C++ Programs	26
2.5	Interpretation of Commands with MATLAB® and Octave		27
2.6	Computer Problem Solving		28
2.7	Summary		30
Key Terms			31
Exercises			31

3 MATLAB and Octave Programming — 33

3.1	Introduction	33
3.2	The MATLAB and Octave Prompt	33
3.3	Variables and Constants	35
3.4	Assignment Statements	35
3.5	Simple Mathematical Expressions	36
3.6	Scientific Notation	37
3.7	Built-In Mathematical Functions	37
3.8	Internal Documentation	38
3.9	Summary	39
Key Terms		39
Exercises		39

II Computational Models — 41

4 Introduction to Computational Models — 43

4.1	Introduction		43
4.2	Preliminary Concepts		43
4.3	A Simple Problem: Temperature Conversion		46
	4.3.1	Initial Problem Statement	46
	4.3.2	Analysis and Conceptual Model	46
	4.3.3	The Mathematical Model	47
4.4	Using MATLAB and Octave		48
	4.4.1	Basic MATLAB and Octave Commands	48
	4.4.2	The Computational Model	48
	4.4.3	Using Data Lists with MATLAB and Octave	50
	4.4.4	Implementation of Model with Data Lists	53
4.5	Summary		55
Key Terms			56
Exercises			56

5 Computational Models and Simulation — 57
- 5.1 Introduction — 57
- 5.2 Categories of Computational Models — 57
- 5.3 Development of Computational Models — 59
- 5.4 Simulation: Basic Concepts — 61
  - 5.4.1 Simulation Models — 62
  - 5.4.2 Simulation Results — 64
- 5.5 Modular Decomposition — 64
- 5.6 Average and Instantaneous Rate of Change — 65
- 5.7 Area under a Curve — 67
- 5.8 The Free-Falling Object — 68
  - 5.8.1 Initial Problem Statement — 69
  - 5.8.2 Analysis and Conceptual Model — 69
    - 5.8.2.1 Assumptions — 70
    - 5.8.2.2 Basic Definitions — 70
  - 5.8.3 The Mathematical Model — 70
  - 5.8.4 The Computational Model — 71
    - 5.8.4.1 Simple Implementation — 71
    - 5.8.4.2 Implementation with Arrays — 75
    - 5.8.4.3 Computing the Rates of Change — 78
- 5.9 Summary — 79
- Key Terms — 80
- Exercises — 80

6 Algorithms and Design Structures — 83
- 6.1 Introduction — 83
- 6.2 Problem Solving — 83
- 6.3 Algorithms — 84
- 6.4 Describing Data — 84
- 6.5 Notations for Describing Algorithms — 85
  - 6.5.1 Flowcharts — 85
  - 6.5.2 Pseudo-Code — 86
- 6.6 Algorithmic Structures — 87
  - 6.6.1 Sequence — 88
  - 6.6.2 Selection — 88
  - 6.6.3 Repetition — 88
- 6.7 Implementation of Algorithms — 90
  - 6.7.1 Programming Languages — 91
  - 6.7.2 Assignment and Arithmetic Expressions — 92
  - 6.7.3 Simple Numeric Computations — 92
  - 6.7.4 Simple Input/Output — 93
- 6.8 Computing Area and Circumference — 95
  - 6.8.1 Specification — 95

viii

		6.8.2	Algorithm with the Mathematical Model	95
	6.9	Summary		97
	Key Terms			97
	Exercises			97

7 Selection — 99
- 7.1 Introduction . . . 99
- 7.2 Selection Structure . . . 99
  - 7.2.1 General Concepts of the Selection Structure . . . 99
  - 7.2.2 Selection with Pseudo-Code . . . 100
  - 7.2.3 Selection with MATLAB and Octave . . . 101
  - 7.2.4 Conditional Expressions . . . 101
  - 7.2.5 Example with Selection . . . 102
- 7.3 Complex Numbers with MATLAB and Octave . . . 104
- 7.4 A Computational Model with Selection . . . 106
  - 7.4.1 Analysis and Mathematical Model . . . 106
  - 7.4.2 Algorithm for General Solution . . . 106
  - 7.4.3 Detailed Algorithm . . . 107
- 7.5 Multilevel Selection . . . 109
  - 7.5.1 General Multipath Selection . . . 110
  - 7.5.2 The Case Structure . . . 111
- 7.6 Complex Conditions . . . 113
- 7.7 Summary . . . 114
- Key Terms . . . 114
- Exercises . . . 115

8 Repetition — 117
- 8.1 Introduction . . . 117
- 8.2 Repetition with While Construct . . . 117
  - 8.2.1 While-Loop Flowchart . . . 118
  - 8.2.2 The While Structure in Pseudo-Code . . . 118
  - 8.2.3 While-Loop with MATLAB and Octave . . . 119
  - 8.2.4 Loop Counter . . . 119
  - 8.2.5 Accumulator Variables . . . 120
  - 8.2.6 Summation of Input Numbers . . . 121
- 8.3 Repeat-Until Construct . . . 122
- 8.4 For Loop Structure . . . 124
  - 8.4.1 The Summation Problem with a For Loop . . . 125
  - 8.4.2 The Factorial Problem . . . 126
    - 8.4.2.1 Mathematical Specification of Factorial . . . 126
    - 8.4.2.2 Computing Factorial . . . 126
- 8.5 Summary . . . 127
- Key Terms . . . 128

Exercises 128

9 Data Lists **131**
9.1 Introduction 131
9.2 Creating an Array 132
9.2.1 Creating Arrays in Pseudo-Code 132
9.2.2 Creating Arrays in MATLAB and Octave 133
9.3 Operations on Arrays 135
9.3.1 Array Elements in Pseudo-Code 136
9.3.2 Using Array Elements with MATLAB and Octave . . 136
9.3.3 Arithmetic Operations on Vectors 137
9.4 Multidimensional Arrays 138
9.4.1 Multidimensional Arrays with Pseudo-Code 138
9.4.2 Multidimensional Arrays with MATLAB and Octave . 139
9.5 Applications Using Arrays 141
9.5.1 Problems with Simple Array Manipulation 141
9.5.1.1 The Average Value in an Array 141
9.5.1.2 Maximum Value in an Array 143
9.5.2 Searching 145
9.5.2.1 Linear Search 145
9.5.2.2 Binary Search 147
9.6 Average and Instantaneous Rate of Change 150
9.6.1 Average Rate of Change 150
9.6.2 Instantaneous Rate of Change 151
9.6.3 Computing the Rates of Change 152
9.7 Area under a Curve 156
9.8 Summary 159
Key Terms 160
Exercises 160

10 Modules **163**
10.1 Introduction 163
10.2 Modular Design 163
10.3 MATLAB and Octave Script Files 165
10.4 Functions 166
10.4.1 Function Definition 166
10.4.2 Function Definition in MATLAB and Octave 167
10.4.3 Simple Function Calls 168
10.4.4 Functions with Parameters 169
10.4.5 Function Calls with Data 170
10.4.6 Functions with Return Data 171
10.5 Documenting MATLAB and Octave Functions 172
10.6 Summary 173

| Key Terms . . . . | 173 |
| Exercises . . . . | 174 |

11 Mathematical Models: Basic Concepts — 175
- 11.1 Introduction 175
- 11.2 From the Real-World to the Abstract World 175
- 11.3 Discrete and Continuous Models 177
- 11.4 Difference Equations and Data Lists 177
- 11.5 Functional Equations 180
- 11.6 Validating a Model 181
- 11.7 Models with Arithmetic Growth 181
- 11.8 Using MATLAB and Octave to Implement the Model 182
    - 11.8.1 MATLAB and Octave Implementation 182
    - 11.8.2 Producing the Charts of the Model 183
- 11.9 Summary 185
- Key Terms 185
- Exercises 186

12 Models with Quadratic Growth — 187
- 12.1 Introduction 187
- 12.2 Quadratic Growth 187
- 12.3 Differences of the Data 188
- 12.4 Difference Equations 191
- 12.5 Functional Equations 192
- 12.6 Models with Quadratic Growth 193
    - 12.6.1 Simple Quadratic Growth Models 193
    - 12.6.2 Models with Sums of Arithmetic Growth 194
- 12.7 Solution and Graphs of Quadratic Equations 196
- 12.8 Summary 198
- Key Terms 198
- Exercises 198

13 Models with Polynomial Functions — 201
- 13.1 Introduction 201
- 13.2 General Forms of Polynomial Functions 201
- 13.3 Evaluation and Graphs of Polynomial Functions 202
    - 13.3.1 Evaluating Polynomial Functions 202
    - 13.3.2 Generating Graphs of Polynomial Functions 205
- 13.4 Solution to Polynomial Equations 206
- 13.5 Summary 209
- Key Terms 209
- Exercises 209

14 Data Estimation and Empirical Modeling — 211
- 14.1 Introduction 211
- 14.2 Interpolation 211
    - 14.2.1 Linear Interpolation 212
    - 14.2.2 Nonlinear Interpolation 215
- 14.3 Curve Fitting 218
- 14.4 Summary 220
- Key Terms 221
- Exercises 221

15 Models with Geometric Growth — 225
- 15.1 Introduction 225
- 15.2 Basic Concepts of Geometric Growth 225
    - 15.2.1 Geometric Growth with Increasing Data 226
    - 15.2.2 Geometric Growth with Decreasing Data 226
    - 15.2.3 Geometric Growth: Case Study 1 227
    - 15.2.4 Geometric Growth: Case Study 2 230
- 15.3 Functional Equations in Geometric Growth 232
- 15.4 Properties of Exponential Functions 233
    - 15.4.1 Exponentiation 233
    - 15.4.2 Logarithms 233
- 15.5 Summary 236
- Key Terms 236
- Exercises 236

16 Vectors and Matrices — 239
- 16.1 Introduction 239
- 16.2 Vectors 240
- 16.3 Simple Vector Operations 240
    - 16.3.1 Arithmetic Operations 241
    - 16.3.2 Applying Vector Functions 243
- 16.4 Matrices 245
    - 16.4.1 Arithmetic Operations 246
    - 16.4.2 Function Application 247
- 16.5 Array Indexing 247
- 16.6 Plotting Vectors 249
- 16.7 Summary 251
- Key Terms 251
- Exercises 251

17 Text Data **253**
  17.1 Introduction 253
  17.2 String Vectors 253
    17.2.1 String Operations 253
    17.2.2 String Functions 254
  17.3 String Matrices 257
  17.4 Summary 258
  Key Terms 258
  Exercises 258

18 Advanced Data Structures **259**
  18.1 Introduction 259
  18.2 Cell Arrays 259
  18.3 Structures 264
  18.4 Summary 268
  Key Terms 268
  Exercises 268

Appendix A MATLAB and GNU Octave Software **271**
  A.1 Introduction 271
  A.2 The MATLAB Components 271
  A.3 The Desktop 272
  A.4 Starting MATLAB 274
  A.5 Exiting MATLAB 274
  A.6 The Command Window 275
  A.7 Current User Folder 275
  A.8 The Startup Folder 276
  A.9 Using Command Files (Scripts) 276
  A.10 MATLAB Functions 280
  A.11 GNU Octave 285

Appendix B Computer Systems **289**
  B.1 Introduction 289
  B.2 Computer Systems 289
    B.2.1 Hardware Components 290
      B.2.1.1 Processors 290
      B.2.1.2 Main Memory 291
      B.2.1.3 Storage Devices 291
      B.2.1.4 Input Devices 292
      B.2.1.5 Output Devices 292
      B.2.1.6 Bus 292
    B.2.2 Computer Networks 292
    B.2.3 Software Components 293

	B.3	Operating Systems	. . . .	294
		B.3.1 Operating System User Interfaces	294
		B.3.2 Contemporary Operating Systems	295
		B.3.2.1 Unix	295
		B.3.2.2 Microsoft Windows	296
B.4	Summary			296
Key Terms				296

Bibliography   **297**

Index   **299**

List of Figures

1.1	Identifying nouns and verbs.	4
1.2	Area of rectangle.	4
1.3	Formula from nouns and verbs.	5
1.4	Heating water.	5
1.5	Multiple steps problem.	7
1.6	Diagram of counting example.	7
1.7	Diagram with average.	8
1.8	Validating conditions.	11
1.9	Simple data validation.	12
1.10	Multiple conditions.	12
1.11	OR used with one value.	13
1.12	OR used with multiple values.	13
2.1	General structure of a program.	22
2.2	Compiling a Java source program.	26
2.3	Executing a Java program.	26
2.4	Compiling a C++ program.	27
2.5	Linking a C++ program.	27
2.6	MATLAB/Octave interpreter.	28
2.7	The waterfall model.	29
3.1	MATLAB/Octave interpreter.	33
3.2	Simple commands in an Octave window.	34
4.1	Computational science as an integration of several disciplines.	45
4.2	Simple commands in an Octave window.	49
4.3	Plot of the temperature conversion.	54
5.1	Discrete changes of number of cars in the queue.	58
5.2	Development of computational models.	59
5.3	Model development and abstract levels.	61
5.4	High-level view of a simulation model.	63
5.5	The slope of a line.	65
5.6	The slope of a secant.	66

5.7	The slope of a tangent.	67
5.8	The area under a curve.	69
5.9	Computing the height of the falling object in Octave.	73
5.10	Plot of the values of height with time.	77
5.11	Plot of the values of velocity with time.	79
6.1	Transformation applied to the input data.	84
6.2	Simple flowchart symbols.	86
6.3	A simple flowchart example.	87
6.4	A flowchart with a sequence.	88
6.5	Selection structure in flowchart form.	89
6.6	An example of the selection structure.	89
6.7	While loop of the repetition structure.	90
6.8	Repeat-until loop of the repetition structure.	91
6.9	Flowchart data input/output symbol.	94
7.1	Flowchart of the selection structure.	100
7.2	Example of selection structure.	103
7.3	Complex number P in the complex plane.	104
7.4	High-level flowchart for the quadratic equation.	107
7.5	Solving the quadratic equation in Octave.	109
8.1	A flowchart with a while-loop	118
8.2	A flowchart with a repeat-until structure	123
9.1	A simple array.	131
9.2	A simple two-dimensional array.	132
9.3	The slope of a line.	151
9.4	The slope of a secant.	152
9.5	The slope of a tangent.	153
9.6	The velocity of the free-falling object.	154
9.7	Bar chart of the velocity of the free-falling object.	155
9.8	The area under a curve.	157
10.1	Modular structure of a computational model.	164
10.2	Module communication and interface.	164
10.3	A simple function call.	168
11.1	Real-world to abstract-world mapping.	176
11.2	Discrete model.	178
11.3	Continuous model.	178
11.4	Monthly price of electric energy.	184
11.5	Monthly price given and calculated of electric energy.	185

xvii

12.1	Number of patients for 1995–2002.	188
12.2	Original data.	189
12.3	Original data and differences.	190
12.4	Number of links to connect n computers.	195
12.5	Graph of a quadratic equation.	197
13.1	Graph of the equation $y = 2x^3 - 3x^2 - 36x + 14$.	205
13.2	Graph of the equation $y = 3x^5 - 2$.	206
13.3	Graph of equation $y = 23x^4 - 17x^3 - 14x^2 + 3x + 3$.	207
14.1	Graph of linear interpolation of an intermediate data point.	212
14.2	Graph of linear interpolation of multiple intermediate data points.	215
14.3	Graph of given data points.	216
14.4	Graph of given and estimated data points.	217
14.5	Graph of fitted linear polynomial.	219
14.6	Regression of linear polynomial.	220
14.7	Regression of a polynomial of degree 3.	221
15.1	Data with geometric growth.	226
15.2	Data decreasing with geometric growth.	227
15.3	Population of a small town for 1995–2003.	229
15.4	Impurities in water (parts/gallon).	231
15.5	A typical exponential function, $95.25 \cdot 1.6^t$.	234
15.6	Natural logarithm.	235
16.1	A simple plot.	250
18.1	A simple cell array.	261
A.1	MATLAB desktop.	273
A.2	MATLAB Command Window.	275
A.3	MATLAB Current Folder.	276
A.4	MATLAB Editor.	277
A.5	MATLAB Current Folder.	277
A.6	The Octave window.	286
A.7	Octave help documentation on *plot*.	287
A.8	Octave documentation manual.	287
B.1	Basic hardware structure of a computer system.	291
B.2	Basic structure of a network.	293

List of Tables

4.1 Celsius and Fahrenheit temperatures. 50

5.1 Values of height and vertical velocity. 74
5.2 Values of the rates of change of height and vertical velocity of the free-falling object. 80

11.1 Average price of electricity (cents per kW-h) in 2010. 180

12.1 Number of patients for years 1995–2002. 187
12.2 Number of cable installations for years 1995–2002. 195

15.1 Population of a small town during 1995–2003 (in thousands). . 227

Preface

Computational science is an emerging area (or discipline) that includes concepts, principles, and methods from applied mathematics and algorithmic design and computer programming; these are applied in various areas of science and engineering to solve large-scale scientific problems. A *computational model* is a computer implementation of the solution to a (scientific) problem for which a mathematical representation has been formulated. Developing a computational model includes formulating the mathematical representation and implementing it by applying computer science concepts, principles and methods. Computational modeling is the foundational component of computational science and focuses on reasoning about problems using *computational thinking* and developing models for problem solving.

The primary goal of this book is to introduce readers to the basic principles of computational modeling at the level of fundamental concepts. Emphasis is on reasoning about problems, conceptualizing the problem, elementary mathematical modeling, and their computational solution that involves computing results and visualization.

The book emphasizes analytical skill development and problem solving rather than programming language syntax and explains simple pseudo-code, simple algorithm design, and MATLAB®, Octave, and FreeMat programming. MATLAB is a widely used scientific software tool; Octave and FreeMat are freely available software tools compatible with MATLAB.

MATLAB® is a registered trademark of The Mathworks, Inc. For product information, please contact: The MathWorks, Inc.

3 Apple Hill Drive
Natick, MA 01760-2098 USA
Tel: 508 647 7000
Fax: 508-647-7001
E-mail: info@mathworks.com
Web: www.mathworks.com

Although computational models require high-performance computing to solve large and complex scientific problems, this book presents only the elementary notions of computational models. For beginning undergraduate students (of science and engineering), the goal is to provide relevant material for easy understanding of computational models and their development, as early as possible in their studies. Therefore, the book takes an "early introduction" approach to computational models by providing the readers with a strong foundation in computational modeling by applying elementary mathematical concepts and basic principles, methods, and techniques of computer science.

The book consists of two parts. Part I presents the basic concepts and principles of computational models. In this part, the book discusses basic modeling and techniques for designing and implementing problem solutions, independent of software and hardware tools. Standard pseudo-code constructs and flowcharts are explained and applied in designing models for various case studies. These are implemented using only the minimum necessary knowledge of MATLAB, a widely used scientific software tool, and Octave, a freely available tool that is compatible with MATLAB. Part II of the book presents the elementary mathematical modeling principles and more detailed computer implementation of models with programming constructs using MATLAB and Octave. Examples and case studies demonstrate the computation and visualization of computational models.

The basic syntax constructs of MATLAB and Octave are presented gradually and the various programming principles are explained in an incremental manner for the actual implementation of computational models. This book avoids presenting too much syntax of the programming language at the beginning, which usually results in unnecessary difficulty for the student in understanding the underlying concepts in problem solution and programming.

There are very few books on the basic treatment of computational models; this book and its associated models were designed as teaching materials. The book is aimed at beginning college students in computer science, mathematics, statistics, science, and engineering. The main features of the book are the following:

- Explanation of the basic concepts and principles of computational models are explained.

- Discussion of the models is based on elementary mathematical models, at the level of pre-calculus.

- Emphasis on modularity and abstraction that help in dealing with large and complex models. These concepts are introduced early in Chapters 1 and 2.

- Discussion of the design of algorithmic solutions to problems using standard flowcharts and pseudo-code.

- Implementation of the problem solutions and the corresponding models in MATLAB and Octave.

The material in this book is adequate at the beginning undergraduate level, a CS0 course or a CS1 course. As mentioned previously, the material can also be used in other undergraduate curricula, e.g., mathematics, engineering, and the other sciences. Because of the level of mathematics used, the book can also be utilized for a course in college preparation at the high school level.

This book will be extremely helpful to readers who have never programmed before and the book can help improve the overall approach in teaching/learning programming principles by emphasizing more problem solving, abstraction, algorithm design with pseudo-code, and basic programming before teaching/learning more advanced programming principles with standard programming languages such as Java, C++, Ada, and others.

I acknowledge the help provided through discussions with some of my colleagues, namely Dr. Ben Setzer and Dr. Dick Gayler. I also acknowledge and give special thanks to Professor Rich Schlesinger for writing Chapter 1.

José M. Garrido
Kennesaw, Georgia

About the Author

José M. Garrido is professor of computer science in the Department of Computer Science, Kennesaw State University, Georgia. He holds a Ph.D. from George Mason University in Fairfax, Virginia, an M.S.C.S also from George Mason University, an M.Sc. from the University of London, and B.S. in electrical engineering from the Universidad de Oriente, Venezuela.

Dr. Garrido's research interest is on object-oriented modeling and simulation, multi-disciplinary computational modeling, formal specification of real-time systems, language design and processors, modeling systems performance, and software security. Dr. Garrido developed the Psim3, PsimJ, and PsimJ2 simulation packages for C++ and Java. He has recently developed the OOSimL, the Object-Oriented Simulation Language (with partial support from NSF).

Dr. Garrido has published several papers on modeling and simulation, and on programming methods. He has also published six textbooks on object-oriented simulation and operating systems.

Part I

Understanding Problem Solving

Chapter 1

Understanding Word Problems

1.1 Introduction

Although a computational problem may be solved using mathematics, a calculator, or a computer program, developing a solution to these problems begins with a description of the problem in a human language (e.g. English). To solve such problems, it is necessary to understand what the (English) problem statement is saying. One may say "well, I already understand English," and that is correct — for casual conversations. In this case, one only needs to have a general **understanding** of what is being said.

The problem statement of real-world situations is often described in a very ambiguous, incomplete, and confusing manner. However, for developing computation models from problem statements, one needs to understand *precisely* what is being said. Therefore, there is a need to focus on *every word* that is being said (not just the general context). Otherwise, the appropriate computations will most likely not be performed.

The first important step in developing computational models is understanding the problem. This is part of the general approach of computational thinking that is required for any problem that involves calculations. This chapter discusses how to understand a precise problem description.

Computational thinking is an approach for developing computational models and is used for problem solving in the following application areas:

- Sciences (biology, chemistry, computing, physics, geology, etc.)

- Social Sciences (psychology, sociology, geography, etc.)

- Engineering (electrical engineering, civil engineering, mechanical engineering, etc.)

- Business (accounting, finance, marketing, economics, risk management, etc.)

1.2 Nouns and Verbs

Consider the following example:

To calculate the area of a rectangle, multiply the width and height.

This problem is so simple that our minds may automatically grasp what to do. Other problems are not so simple. So, let's develop a technique for understanding a problem like this and then use that technique on more complex problems. The key to understanding a word problem is to find the nouns and the verbs. As shown in Figure 1.1, the nouns in this problem are "area," "width," and "height." The verbs are "calculate" and "multiply."

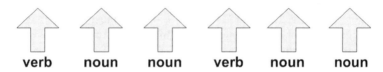

FIGURE 1.1: Identifying nouns and verbs.

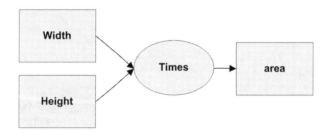

FIGURE 1.2: Area of rectangle.

The nouns identify *data values* to use. The nouns also give us names for those values. The verbs describe what *actions* are to be performed.

This is illustrated in Figure 1.2. In this type of diagram, the blocks indicate steps in the solution and the arrows show the **uses** of the data. This diagram illustrates the concepts of **uses** and **produces**. It shows that width and height will be **used by** the times operation, which then **produces** the value of area. This is a key concept. Once we know what the nouns and verbs are, we can understand the computation by understanding how the values we have are **used** and what values are **produced**. Once we understand that, then it is easy to create a formula for the problem.

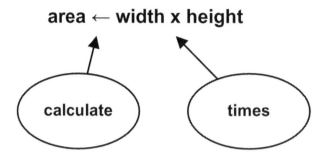

FIGURE 1.3: Formula from nouns and verbs.

We can then reason that this word problem becomes the formula in Figure 1.3. Now consider a more complicated example:

> **Calculate** the energy needed to **heat** water from an initial temperature to a final temperature. The formula to **compute** the energy is the amount of water in kilograms **times** the **difference** of the final and initial temperatures in Celsius times 4184.

The nouns are: "energy," "water," "initial temperature," "final temperature," "amount," "difference,"and "formula." The verbs are "calculate," "heat," "compute,'' and "times."

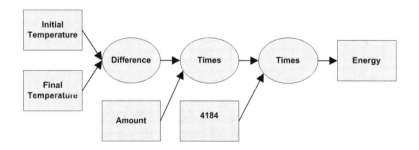

FIGURE 1.4: Heating water.

All of the nouns listed refer to values except for "water." This noun helps us understand the physical context of the problem, but it can be ignored for the calculations. Similarly, all of the verbs refer to a computational action except "heat." Again, this verb helps us understand the physical context of the problem, but we can ignore it for the computation.

The word "difference" is special. It is a noun, indicating it describes a value, but the meaning of that word also describes an action (subtraction).

Now we can draw the diagram in Figure 1.4. This clearly shows how the values are used and what values are produced. In this type of diagram, the blocks indicate steps in the solution and the arrows show the **uses** of the data. From that diagram we can then develop the formula for this problem.

$$Energy = amount \times (final\ temp - initial\ temp) \times 4184$$

1.3 Problem Decomposition

Many problems cannot be solved with a single equation. These types of problems need to be decomposed into smaller problems. We solve the smaller problems and then put the pieces together to get the solution to the entire problem.

Consider the problem discussed previously, but with an additional step:

Calculate the energy needed to heat water from an initial temperature to a final temperature. The user will provide the amount of water in kilograms and the initial and final temperatures of the water in Celsius. The formula to compute the energy is the amount of water in kilograms times the difference of the final and initial temperatures times 4184.

The 2nd sentence describes actions to be performed by a person before we can do the computation. Thus, this problem has two steps:

1. User Input

2. Calculate Energy

This kind of multiple step problem is best visualized in a diagram like the one shown in Figure 1.5. The figure illustrates the idea of **uses**. First, the user provides data. Then, the calculation **uses** that data. In this type of diagram, the blocks indicate steps in the solution and the arrows show the **uses** of the data. Each of the boxes in this diagram represents a more detailed diagram that describes the actual steps to solve that part of the whole problem. Now, consider a more complicated problem:

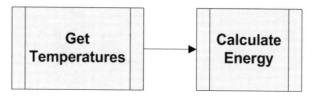

FIGURE 1.5: Multiple steps problem.

Count how many numbers are greater than the average

When we analyze this problem, we discover the nouns are "numbers" and "average." The verb is "count." Thus, this calculation **uses** two values (numbers and average). This is illustrated in Figure 1.6.

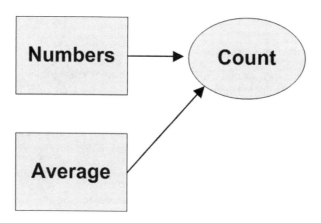

FIGURE 1.6: Diagram of counting example.

However, the average is not a separate independent value, it must be calculated. To calculate the average, we need to **use** numbers. This is shown in Figure 1.7.

This diagram is important because it visually indicates:

- What calculations use what values

- The order in which the calculations must occur.

The first calculation uses "average." Consequently, we must calculate "average" before we perform that step.

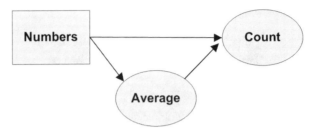

FIGURE 1.7: Diagram with average.

1.4 The Language of Computational Problems

The English problem statement may use certain special terms.

1.4.1 Terms for Computed Values

One type of such special terms are nouns that refer to a value and an implied calculation to obtain that value. All of these special nouns are usually followed by the word **of** and then the value(s) to be in the implied calculation.

- As seen previously, the word **difference** is both a noun which refers to a value and the result of a calculation (subtraction).

- The word **inverse** is a noun but also refers to the result of either:

    - Taking a fraction and reversing the numerator and denominator. The inverse of $\frac{2}{5}$ is $\frac{5}{2}$.
    - Dividing 1 by a specified whole number. The inverse of 5 is $\frac{1}{5}$.

- The word **sum** or **total** refers to a value which is the addition of potentially many values. These values may be explicitly stated or a range of values may be specified.

    The sum of all the weights

- The word **product** refers to a value which is the multiplication of several values. These values will usually be explicitly stated.

    The product of the width and height

- The word **negative** can either be an adjective preceding some value or it can be a noun.

- negative total
- the negative of the total

- The word **average** refers to the sum of many values divided by the count of how many values there are.

The average of all the scores

- The word **number** or **count** often refers to how many items of a certain type there are.

The number of values greater than the average

1.4.2 Implied Phrases

Often, an English statement will not explicitly include a phrase. Instead, it will be implied by the context. Consider this problem:

To compute the average of a set of numbers, calculate the sum and divide by how many numbers there are

This problem statement is missing "of all of the numbers" immediately after the word "sum." The missing phrase is implied by the context of the entire statement. Thus, it is important to look for missing phrases like this one and add them to the problem statement to fully understand the problem.

1.4.3 Units of Measure

Problem statements often refer to "units of measure" (e.g. kilograms, days, meters, etc.). To add or subtract two values, they must be measures of the same type of thing. You can add or subtract two distances, but you cannot add a distance and a weight. If two values measure the same type of thing but are expressed in different units of measure, you must first convert them to the same unit of measure in order to add or subtract them. For example, if you need to add 5 inches and 2 meters, you must either convert the inches to meters or the meters to inches.

Units of measure often have the word "per" (e.g. kilometers per hour, miles per gallon, etc.). Whenever you see the word "per," this is a description of a **ratio** or **rate**. The first step to solving this type of problem is to rewrite each "ratio" as a fraction.

$$5 \text{ miles per hour} = \frac{5 \text{ miles}}{\text{hour}}$$

Problems that refer to ratios will often require multiplication. When you multiply units of measure, a unit in a numerator will cancel out the same unit in a denominator. So, for this problem,

At 5 miles per hour, how far will you go in 10 hours?

We would use this formula.

$$\frac{5 \text{ miles}}{\text{hour}} \times 10 \text{ hours}$$

The hours cancel each other

$$\frac{5 \text{ miles}}{\cancel{\text{hour}}} \times 10 \cancel{\text{hours}}$$

and we are left with 5 miles x 10. The answer will be 50 miles. Sometimes, this type of problem also requires converting units of measure. Suppose you have this problem:

At 5 miles per hour, how far will you go in 30 minutes?

We use this formula:

$$\frac{5 \text{ miles}}{\text{hour}} \times 30 \text{ minutes}$$

Hours and minutes are not the same and do not cancel. So, we must include a conversion ratio

$$\frac{5 \text{ miles}}{\text{hour}} \times 30 \text{ minutes} \times \frac{1 \text{ hour}}{60 \text{ minutes}}$$

Now, the hours will cancel and the minutes will cancel

$$\frac{5 \text{ miles}}{\cancel{\text{hour}}} \times 30 \cancel{\text{minutes}} \times \frac{1 \cancel{\text{hour}}}{60 \cancel{\text{minutes}}}$$

The result is 2.5 miles. Thus, to solve this type of problem, you need to construct a formula that will:

- Result in all of the units being canceled except the one for the desired answer.

- Have the remaining uncanceled (desired) unit in the numerator.

To construct a formula, you can invert units of measure, as necessary. So, for this problem:

At 5 miles per hour, how long will it take to go 10 miles?

Understanding Word Problems

Our answer should be a measure of time (hours). So, we invert our miles per hour ratio to get this formula:

$$\frac{\text{hour}}{5 \text{ miles}} \times 10 \text{ miles}$$

In this formula, the miles cancel and we have

$$\frac{\text{hour}}{5} \times 10$$

which yields 2 hours.

1.4.4 Conditions

A common occurrence in computations is a statement that the computation is valid only if certain conditions apply. For example, the area of a square is valid only if the length of a side is greater than 0. This is illustrated in Figure 1.8. The details of the data validation are shown in Figure 1.9.

FIGURE 1.8: Validating conditions.

We describe a condition using an If statement that precedes the calculation. Thus, the above condition is written as:

IF side > 0, area = side × side

Sometimes, there will be multiple conditions to consider. These multiple conditions will be separated by one of the following:

- The word "and" means that **both** or **all** of the associated conditions must be true. For example:

    The area of a rectangle is length times width if both are greater than 0.

Figure 1.10 illustrates this set of conditions and is written as:

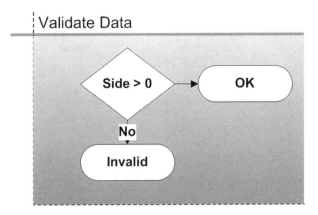

FIGURE 1.9: Simple data validation.

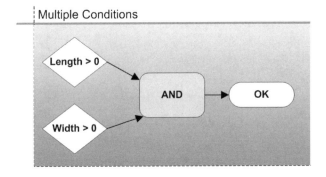

FIGURE 1.10: Multiple conditions.

IF length > 0 AND width > 0, area $=$ length $\times$ width

- In English, the word "or" has different meanings depending on its context. A named value can only have one actual value at any instant. Thus, when "or" is used to refer to the same named value, it means that only one of those possible conditions can occur.

    if distance is less than 100 or greater than 200, ...

On the other hand, when the word "or" is used to refer to values of different named items, it means that either one is fine.

Understanding Word Problems 13

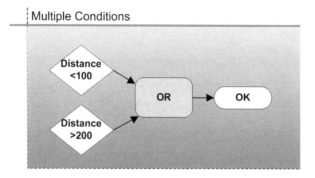

FIGURE 1.11: OR used with one value.

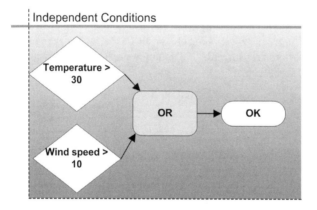

FIGURE 1.12: OR used with multiple values.

if temperature is less than 30 or wind speed greater than 10,
...

1.4.5 Repetition

We often have to repeat the same calculation, but with different values. Any of the following phrases will be an indication that this is required.

- **for.** This will specify a range of values and the calculation(s) to perform on each of the values in the range. An example would be:

    For each value in the range $5 < x < 10$, compute the sum of the inverses.

- **while.** This will specify a condition and the calculation(s) to repeatedly perform as long as the condition is true.

  While the data values are positive, compute their sum

- **until.** This is the opposite of while. "Until" will specify a condition and the calculation(s) to repeatedly perform until the condition is true.

  Until the data values are negative, compute their sum

- **all** or **every.** This will typically specify a calculation to perform on all or every value. An example is:

  Compute the sum of all of the values less than 100.

A problem that involves repetition will often also have parts that do not require repetition. It is important to not confuse these. If you do not repeat parts that need repetition, you will either not use all of the values or will do incorrect calculations. If you repeat parts that do not require repetition, you will typically produce incorrect results.

1.5 Objects

Word problems usually refer to some real-world scenario. Consequently, they refer to one or more physical objects. Certain simple objects have a value associated with them.

- A dime is 10 cents.

A slightly more complicated object will have multiple values associated with it.

- A tank or chemistry beaker will have a capacity and current volume.
- An item being sold will have a sales price and a cost.

Still more complicated objects will have one or more values that depend on a physical action. This is known as the **state** of the object.

- A coin will be heads or tails after it is flipped.
- A dice will show a value after it is rolled.
- An electrical motor will have a rotational speed that depends on the amount of electricity it is receiving.

The problem about heating water is a problem about a physical action (heating) that changes the state of an object (the water) and the resulting change of a value associated with the object (the water's temperature).

Many real-world problems are best understood by understanding the change in state of the object(s) in the problem. We shall look at this in more detail later.

1.6 Problems with Many Numbers

If we need to develop a program to solve a computational problem, we need to understand how we would do it ourselves, using pencil and paper. Often, we can easily say how we do it. Other times, what we do is so implicit, we don't think about how we do it. In those situations, we need to take some effort to understand the problem. Let's consider an example:

How do we determine if a list of numbers is all even?

Our first response will be "Well, we just look at the list". And that is correct (at least for a small list). But, how do we know all the numbers are even? What are we "looking" for? In this particular case, there is a *pattern* that the numbers form. That pattern is determined by the meaning of the word *even*. That word means that when we divide a number by 2, the remainder is 0. Thus, we look at each number to see if that condition holds. Several important points:

- If we find a number that is not even, do we keep looking at the rest of the numbers? No! We stop immediately and report that the list is not all even.

- When can we say the list is all even? Only after we have looked at all of the numbers!

Let's consider another example:

How do we determine if a list of numbers is sorted (in ascending order)?

This is very similar to the previous problem. We are looking at the numbers for a particular pattern. That pattern is determined by the meaning of the word *sorted*. That word means that each number is $\geq$ the previous number (or each number is $\leq$ the next number). Thus, we begin by comparing the first number to the next and then compare the 2nd number to the one after it and so on. Several important points can be noted:

- Just like the previous example if we find a pair of numbers that are out of order, do we keep comparing? No! We stop immediately and report that the list is not sorted.

- When can we say the list is sorted? Only after we have looked at all of the pairs!

- Is there a number after the last number? Of course not. But that means that our last comparison will not be looking at the last number and the one after it. Instead, the last comparison will be looking at the next to last number and the one after it.

These examples illustrate that the key to understanding how we solve a problem is to determine

- The *pattern* that the problem presents.

- How we deal with that pattern

1.6.1 Working with Patterns

Now, let's consider another example that is similar to our first example:

How do we determine if a list of numbers has any that are even?

Once again, we are looking for a pattern. It's the same pattern from the first example, but when we stop is different. Several important points:

- If we find a number that is even do we keep looking at the rest of the numbers? No! We stop immediately and report that the list has an even number.

- When can we say the list has no even numbers? Only after we have looked at all of the numbers!

All of these examples involve examining a list of values to see if some condition holds. The difference in operation between this example and our first example is the key terms *all* and *any*.

- When the condition involves *all*, then you must look at *all* the values before you can say the condition is true, but you can say the condition is not true as soon as you find a value that does not meet the pattern.

- On the other hand, when the condition involves *any*, you can say the condition is true as soon as you find a value that meets the pattern. You must look at all of the values before you can infer that the condition is not true.

Let's consider another example:

$$x = \frac{1}{1} + \frac{1}{3} + \frac{1}{5} + \ldots + \frac{1}{n}$$

The pattern here is quite easy to see. We have the sum of fractions where the denominators increase by 2 each time. Thus, we can see that there are two major types of problems involving patterns.

- Problems that involve computations involving multiple values. These types of problems can usually be expressed in a mathematical equation.

- Problems involving a search to see if a certain pattern exists among the multiple values. This problem is often expressed in words. The solution to this type of problem requires a more explicit effort to understand the pattern involved.

1.7 Summary

In this chapter, three key points have been identified about solving word problems:

- One understands a problem when one can identify the nouns and the verbs. The nouns identify the values to be used in the problem. The verbs identify what type of calculations to perform. We learned that in English, some terms can refer to both a value and how that value is calculated.

- In complex problems, one needs to break the problem into the individual steps identified in the problem and then proceed to analyze and perform each step individually.

- Problems involving multiple values often have a *pattern* that needs to be understood in order to solve the problem.

Key Terms

problem understanding	problem decomposition	patterns
all	and	average
count	difference	every
for	inverse	negative
noun	number	object
or	pattern	sum
total	until	verb
while		

Exercises

For each of the problems below, identify the nouns and verbs and then create a diagram showing the calculations.

Exercise 1.1 Phil has 5 quarters, 3 dimes, and 2 pennies. How much money does he have?

Exercise 1.2 Body mass index (BMI) is a health measure of your weight relative to your height. Your BMI is calculated by taking your weight in kilograms and dividing by the square of your height in meters. How do you calculate your BMI if you know your weight in pounds and your height in inches. One pound is 0.45359237 kilograms. One inch is 0.0254 meters.

Exercise 1.3 The area of a regular hexagon is 3 halves times the square root of 3 times the square of the length of a side.

Exercise 1.4 In physics, acceleration is calculated as the difference between the starting and ending velocities divided by the time.

Exercise 1.5 The length of runway that a plane needs to take off is the square of the takeoff velocity divided by two times the acceleration (see previous exercise).

Exercise 1.6 The Future investment value of an investment amount is the investment amount times one plus the monthly interest rate raised to the power of the number of months.

Exercise 1.7 Determine if any of the numbers in a set is odd. Indicate how to solve the problem.

Exercise 1.8 Determine if any value in a set of numbers is within a certain range. Indicate how to solve the problem.

Exercise 1.9 Determine whether the numbers in a set are all within a specified tolerance value of another number. Indicate how to solve the problem.

Exercise 1.10 Determine whether the points in a set are all within a specified circle. Indicate how to solve the problem.

Chapter 2

Problem Solving and Computing

2.1 Introduction

A **model** is a representation of a real system or part of it. **Modeling** is the activity of building models. A **computational model** is a mathematical model implemented in a computer system and usually requires high performance computational resources to execute. The model is used to study the behavior of a complex real system. The computer implementations of computational models are essentially programs.

A computer **program** is a sequence of instructions and data definitions. The instructions allow the computer to manipulate the data to carry out computations and produce desired results when the program executes.

This chapter presents elementary programming concepts that include variables, constants, programming languages, and program execution.

2.2 Programs

A *program* is a *software* component of a computer system and consists of data definitions and instructions that manipulate the data. A program is normally written in a *programming language*. The general structure of a program is shown in Figure 2.1. It consists of.

- Data definitions, which define all the data to be manipulated by the instructions.

- A sequence of instructions, which perform the *computations* on the data in order to produce the desired results.

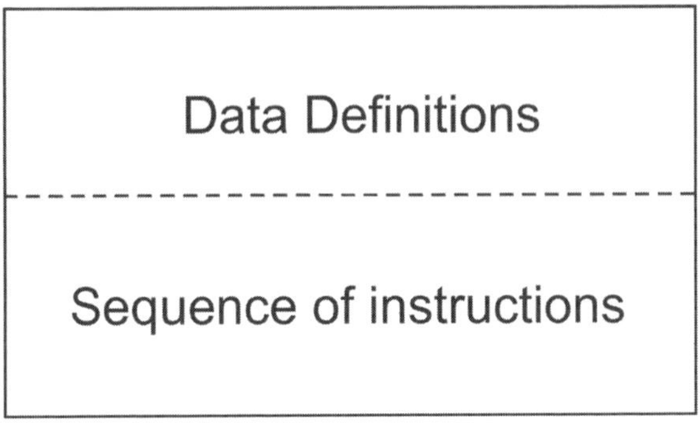

```
┌─────────────────────────────────────┐
│                                     │
│          Data Definitions           │
│                                     │
├─ ─ ─ ─ ─ ─ ─ ─ ─ ─ ─ ─ ─ ─ ─ ─ ─ ─ ─┤
│                                     │
│       Sequence of instructions      │
│                                     │
└─────────────────────────────────────┘
```

FIGURE 2.1: General structure of a program.

2.3 Data Definitions

Data consists of one or more data items. For every computation there is one or more data items (or entities) associated that are to be manipulated or transformed by the computations (computer operations). A computation requires one or more instructions. The definition of a data item is given by:

- The type of the data item
- A unique name to identify the data item
- An optional initial value

The name of a data item is an *identifier* and is given by the programmer; it must be different from any keyword in the programming language. The type of data defines:

- The set of possible values that the data item can take
- The set of possible operations or computations that can be applied to the data item

2.3.1 Name of Data Items

The special text words or symbols that indicate essential parts of a programming language are called *keywords*. These are reserved words and cannot be used for any other purpose. The other symbols used in an algorithm are the ones for identifying the data items and are called *identifiers*. The identifiers are defined by the programmer.

A unique name or label is given to every data item; this name is an identifier. The problem for calculating the area of a triangle uses five data items, *a, b, c, d*, and *circum*.

The data items usually change their values when they are manipulated by the various operations. For example, the following sequence of instructions first gets the value of *t* then adds the value *t* to *tdelay*,

```
read t          // read value of t from keyboard
add t to tdelay
```

The data items named *t* and *tdelay* are called *variables* because their values change when operations are applied on them. Those data items that do not change their values are called *constants*, for example, *MAX_NUM, PI*, etc. These data items are given an initial value that will never change during the program execution.

When a program executes, all the data items used by the various computations are stored in the computer memory, each data item occupying a different memory location.

2.3.2 Data Types

Data types are classified into the three categories:

- Numeric

- Text

- Boolean

The numeric types are further divided into three types, *integer, float,* and *double*. The non-integer types are also known as fractional, which means that the numerical values have a fractional part.

Values of *integer* type are those that are countable to a finite value, for example, age, number of automobiles, number of pages in a book, and so on. Values of type *float* have a decimal point; for example, cost of an item, the height of a building, current temperature in a room, a time interval (period). These values cannot be expressed as integers. Values of type *double* provide more precision than type *float*; for example, the value of the total assets of a corporation.

Text data items are of two basic types: *character* and type *string*. Data items of type *string* consist of a sequence of characters. The values for these two types of data items are text values. The string type is the most common, such as the text value: 'Welcome!'.

The third data type is used for variables whose values can take a truth-value (true or false); these variables are of type *boolean*.

2.3.3 Data Declarations

The data declarations are the data definitions and include the name of every variable or constant with its type. The initial values, if any, for the data items are also included in the data declaration.

In programming languages such as C, C++, and Java, the declaration of variables has the following basic syntactic structure:

⟨ elementary type ⟩ ⟨ variable_name ⟩

The following lines of pseudo-code are examples of data declarations of two constants and three variables of type *integer*, *float*, and *boolean*.

```
constants
      define PI = 3.1416 of type float
      define MAX_NUM = 200 of type integer
variables
      define count of type integer
      define weight of type float
      define busy of type boolean
```

2.4 Programming Languages

A programming language is a notation that programmers use to write programs using a defined set of syntax and semantic rules. The syntax rules describe how to write well-defined sentences. The semantic rules describe the meaning of the sentences.

2.4.1 High-Level Programming Languages

A high-level programming language is a formal notation in which to write instructions to the computer in the form of a program. A programming language helps programmers in the writing of programs for a large family of problems.

High-level programming languages are hardware independent and are problem-oriented (for a given family of problems). These languages allow more readable programs, and are easy to write and maintain. Examples of these languages are Pascal, C, Cobol, FORTRAN, Algol, Ada, Smalltalk, C++, Eiffel, and Java.

Programming languages like C++ and Java can require considerable effort to learn and master. Several newer and experimental, higher-level, object-oriented programming languages have been developed. Each one has a particular goal.

There are several integrated development environments (IDE) designed for numerical and scientific problem solving that have their own programming language. Some of these computational tools are: MATLAB®, Octave, Mathematica, Scilab, Stella, and Maple.

The solution to a problem is implemented in an appropriate programming language. This program is known as the *source program* and is written in a high-level programming language, such as C++, Eiffel, Java, or others.

Once a source program is written, it is translated or converted into an equivalent program in *machine code*, which is the only programming language that the computer can understand. The computer can only execute instructions that are in machine code.

The program executing in the computer usually reads input data from the input device and after carrying out some computations, it writes results to the output device(s).

2.4.2 Interpreters

An interpreter is a program that translates or converts a user program written in a high-level programming language into an intermediate code and immediately executes this code. Examples of interpreters are the ones used for the following languages: MATLAB, Octave, PHP and PERL.

2.4.3 Compilers

A compiler is a program that translates another program written in a programming language into an equivalent program in machine code, which is the only language that the computer accepts for processing.

In addition to *compilation*, an additional step known as *linking* is required before a program can be executed. Examples of programming languages that require compilation and linking are: C, C++, Eiffel, Ada, Fortran. Other programming languages such as Java, require compilation and interpretation.

2.4.4 Compiling and Execution of Java Programs

To compile and execute programs written in the Java programming language, two special programs are required, the compiler and the interpreter.

26   *Introduction to Elementary Computational Modeling*

The Java compiler checks for syntax errors in the source program and translates it into *bytecode*, which is the program in an intermediate form. The Java bytecode is not dependent on any particular platform or computer system. To execute this bytecode, the Java Virtual Machine (JVM), carries out the interpretation of the bytecode.

FIGURE 2.2: Compiling a Java source program.

Figure 2.2 shows what is involved in compilation of a source program in Java. The Java compiler checks for syntax errors in the source program and then translates it into a program in byte-code, which is the program in an intermediate form.

The Java bytecode is not dependent on any particular platform or computer system. This makes the bytecode very portable from one machine to another.

Figure 2.3 shows how to execute a program in bytecode. The Java virtual machine (JVM) carries out the interpretation of the program in byte-code.

FIGURE 2.3: Executing a Java program.

2.4.5   Compiling and Executing C++ Programs

Programs written in C++ must be compiled, linked, and loaded into memory before executing. An executable program file is produced as a result of linking. The libraries are a collection of additional code modules needed by the program. Figure 2.4 illustrates the compilation of a C++ program. Figure 2.5 illustrate the linkage of the program. The executable program is the final form of the program that is produced. Before a program starts to execute in the computer, it must be loaded into the memory of the computer.

FIGURE 2.4: Compiling a C++ program.

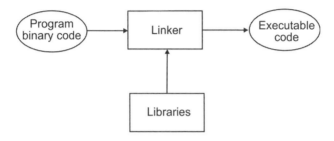

FIGURE 2.5: Linking a C++ program.

2.5 Interpretation of Commands with MATLAB® and Octave

As mentioned previously, MATLAB and Octave are two of several integrated environments designed for numerical and scientific computation and problem solving that have their own programming language. These environments provide an interpreter for their language, which is a program that translates or converts a command written in their high-level programming language and immediately executes the command. A program in MATLAB and Octave is a file of commands, so the program is known as a command file or a *script*.

MATLAB and Octave provide a command window in which all the interaction with a user takes place. The **command prompt** is the symbol that MATLAB and Octave display on the window to alert the user that it is waiting for a command. In MATLAB, the prompt is >>. In Octave, an example of the prompt is octave-3.2.4:n>, in which n is the command line number. Figure 2.6 illustrates the interpretation of a MATLAB and Octave command and the response to the command.

Because MATLAB and Octave have interpreters for their language, there is no direct way to declare the variables and constants used. The following commands are simple assignments to variable y and to variable x. Note how MATLAB and Octave immediately respond to the command.

FIGURE 2.6: MATLAB/Octave interpreter.

```
>> y = 20.6
y =
    20.6000
>> x = 2.56
x =
    2.5600
>> j = 200
j =
    200
```

The MATLAB and Octave interpreter assumes that y is a variable is a non-integer numeric variable since the constant value 20.6 is assigned to y. The same applies to variable x. Variable j is an integer variable because the integer constant 200 is assigned to it. The following command performs a multiplication of the values of variables y and x, the intermediate result is added with constant 125.25, and the final resulting value of this computation is assigned to variable z. When there is no assignment to a variable, MATLAB and Octave display the results with the dummy symbol/variable: ans.

```
>> z = y * x + 125.25
z =
    177.9860
>> z / 12.5
ans =
    14.2389
```

2.6 Computer Problem Solving

Problem solving is the process of producing a computer solution to solve a given real-world problem, it involves the following general tasks:

1. Understanding and describing the problem in a clear and unambiguous form

2. Designing a solution to the problem

3. Developing a computer solution to the problem.

A software development process involves carrying out a sequence of activities. The process is also known as the *software life cycle*.

The simplest model of the software life cycle is the *waterfall model*. This model represents the sequence of phases or activities to develop the software system through installation and maintenance of the software. In this model, the activity in a given phase cannot be started until the activity of the previous phase has been completed.

Figure 2.7 illustrates the sequence of phases that are performed in the waterfall software life cycle.

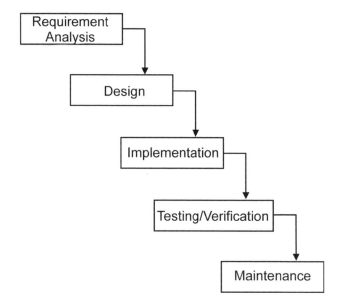

FIGURE 2.7: The waterfall model.

The various phases of the software life cycle are the following:

1. *Analysis*, which results in documenting the problem description and what the problem solution is supposed to accomplish.

2. *Design*, which involves describing and documenting the detailed structure and behavior of the system model.

3. *Implementation* of the software using a programming language.

4. *Testing* and verification of the programs.

5. *Installation*, which results in delivery and installation of the programs.

6. *Maintenance*.

There are some variations of the waterfall model of the life cycle. These include returning to the previous phase when necessary. More recent trends in system development have emphasized an iterative approach, in which previous stages can be revised and enhanced.

A more complete model of the software life cycle is the *spiral model*, which incorporates the construction of *prototypes* in the early stages. A prototype is an early version of the application that does not have all the final characteristics. Other development approaches involve prototyping and rapid application development (RAD).

2.7 Summary

Application programs are programs that the user interacts with to solve particular problems. Computational models are implemented as programs. There are several standard programming languages, such as C, C++, Eiffel, Ada, Java. Compilation is the task of translating a program from its source language to an equivalent program in machine code. Languages in scientific computing environments such as MATLAB and Octave are interpreted.

Computations are carried out on input data by executing individual commands or complete programs.

Key Terms

compilers	linkers	interpreters
application programs	commands	instructions
programming language	Java	C
C++	Eiffel	Ada
bytecode	JVM	program execution
data definition	Source code	high-level language
MATLAB/Octave command	keywords	identifiers
command file	script	program

Exercises

Exercise 2.1   What is a programming language? Why do we need one?

Exercise 2.2   Explain why there are many programming languages.

Exercise 2.3   What are the differences between compilation and interpretation in high-level programming languages?

Exercise 2.4   Explain the purpose of compilation. How many compilers are necessary for a given application? What is the difference between program compilation and program execution? Explain.

Exercise 2.5   What is the real purpose of developing a program? Can we just use a spreadsheet program such as MS Excel to solve numerical problems? Explain.

Exercise 2.6   Find out the differences in compiling and linking Java and C++ programs.

Exercise 2.7   Explain between data definitions and instructions in a program written in a high-level programming language.

Exercise 2.8   For developing small programs, is it still necessary to use a software development process? Explain. What are the main advantages in using a process for program development? What are the disadvantages?

Chapter 3

MATLAB and Octave Programming

3.1 Introduction

This chapter presents a short introduction to the basic principles of programming with MATLAB and Octave. Simple calculations with assignment commands are shown and explained. The use of basic MATLAB and Octave functions is emphasized.

3.2 The MATLAB and Octave Prompt

MATLAB and Octave provide a command window in which all the interaction with a user takes place.

The **command prompt** is the symbol that MATLAB and Octave display on the window to alert the user that it is waiting for a command. In MATLAB, the prompt is >>. In Octave, an example of the prompt is octave-3.2.4:n>, in which n is the command line number. Figure 3.1 illustrates the user interaction with MATLAB and Octave by issuing a command and the response to the command.

FIGURE 3.1: MATLAB/Octave interpreter.

Because MATLAB and Octave use interpreters for their language, there is no direct way to declare the variables and constants. The following commands are examples of simple assignments to variables y, x, and j. Note how MATLAB and Octave immediately respond to the command. Figure 3.2 shows similar commands in an Octave console window.

```
>> y = 20.6
y =
   20.6000
>> x = 2.56
x =
   2.5600
>> j = 200
j =
   200
```

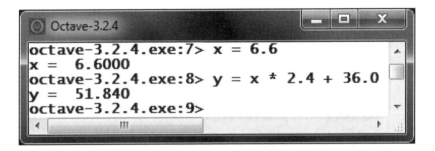

FIGURE 3.2: Simple commands in an Octave window.

The MATLAB and Octave interpreters assume that y is a variable and it is numeric non-integer since the constant value 20.6 is assigned to y. The same applies to variable x. Variable j is an integer variable because the integer constant 200 is assigned to it. The following command performs a multiplication of the values of variables y and x, the intermediate result is added with constant 125.25, and the final resulting value of this computation is assigned to variable z. When there is no assignment to a variable, MATLAB and Octave display the results with the dummy symbol/variable: ans.

```
>> z = y * x + 125.25
z =
   177.9860
>> z / 12.5
ans =
   14.2389
```

A program in MATLAB and Octave is a file of commands, so the program is known as a command file or a *script*.

3.3 Variables and Constants

A **variable** is a data item with a name and value, which will normally change value during execution of a program. A constant is a data item with a name, but its value will not change. MATLAB and Octave provide a few predefined constants. For example, pi, which corresponds to the mathematical quantity π. The predefined value of *pi* is approximately 3.1416. Variables and constants are normally used in an expression. For example:

```
>> y = 0.45 * pi
y =
    1.4137
```

A valid variable name can include any sequence of lower and upper case letters. A few punctuation symbols are allowed. The underscore, _, is the most commonly used non-letter. Numbers are also allowed, at the beginning of the variable name. Spaces are not allowed. Variable names are *case sensitive*, therefore a and A are different variables.

```
A2 = 200.45
MAX_NUM = 200
description = 'T washer'
```

The second assignment command shows a common convention used in programming, a constant name is written all in upper case. The next example is a string assignment. A string value is a sequence of text characters in single quotes.

3.4 Assignment Statements

Giving a value to a variable is known as an assignment and the *assignment operator* is the equals sign, =. The value is given to the variable on the left side of the assignment operator. The right side can be any valid expression. The following MATLAB and Octave command assigns a value to variable *y*. The value is computed in the expression $16.45x + 25.0$, given 0.5 as the value of *x*.

```
>> x = 0.5;
>> y = 16.5 * x + 25.0
y =
    33.2500
```

When an assignment command uses the semicolon at the end, this suppresses the output from a command. MATLAB and Octave perform the assignment, but do not display the value.

3.5 Simple Mathematical Expressions

A basic mathematical **expression** consists of *variables*, *constant* values, and arithmetic *operators*. The variables and constant values in an expression are also known as *operands*. At the prompt, typing an expression causes MATLAB and Octave to **evaluate** the expression and display the result, as seen in the previous example. In general, spaces are used mainly to make expressions more clear and easy to understand.

The basic arithmetic operators are: subtraction that is denoted by a minus sign (−); multiplication by an asterisk (*); and division by a forward slash (/). These operators have been used and shown in previous examples.

The order of operations is: multiplication and division happen before addition and subtraction. Parentheses can be used to override the order of operations.

```
>> z = y * ( x + 125.25 )
z =
   2632.9
```

In addition to the basic arithmetic operators mentioned, exponentiation is another very common operator. MATLAB and Octave use the ^ symbol for exponentiation. The following example shows an expression with variable x raised to the 4th power, for $x = 2.56$.

```
>> y = 4.35 * x^4
y =
   186.8311
```

Exponentiation is performed before multiplication and division, but parentheses can be used to override the order of execution of the operations.

3.6 Scientific Notation

Scientific notation is used by MATLAB and Octave to display very large and very small values. For example:

```
ans = 5.77262e+123
```

The mathematical equivalent for this value is 5.77262×10^{123} and is usually an approximate value. Scientific notation can also be used in assignments or expressions. For example:

```
>> x = 5.4e68
x =
   5.4000e+068
y = x + 126.5e12
y =
   5.4000e+068
```

3.7 Built-In Mathematical Functions

MATLAB and Octave include libraries with built-in functions that compute the most common mathematical functions. An example of these is the library of trigonometric functions. The following command has an expression with the *sin()* function applied to variable *x*. The *argument* of the function call is written in parentheses and its value is assumed in radians.

```
>> y = 2.16 + sin(x)
y =
    2.7094
>> j = 0.335
j =
    0.3350
>> z = x * sin(j * pi)
z =
    0.4343
```

To compute the square root, MATLAB and Octave provide the *sqrt()* function. The following example computes the value of *z* using the mathematical expression:

$$z = \sqrt{\sin^2 x + \cos^2 y}$$

MATLAB and Octave commands that implement the evaluation of the previous expression includes the use of three functions: *sqrt()*, *sin()*, and *cos()*.

```
>> x = 0.5
x =
    0.5000
>> y = 0.8
y =
    0.8000
>> z = sqrt(sin(x)^2 + cos(y)^2)
z =
    0.8457
```

The exponential function *exp()* computes *e* raised to the given power. The following command computes $q = y + x \times e^k$.

```
>> k = 1.685
k =
    1.6850
>> q = y + x * exp(k)
q =
    3.4962
```

To compute the logarithm base *e* of *x*, denoted mathematically as $\ln x$ or $\log_e x$, MATLAB and Octave provide function *log()*.

```
>> r = (q-y)/x
r =
    5.3925
>> t = log(r)
t =
    1.6850
```

3.8 Internal Documentation

It is useful to include *comments* in the command lines. These provide clarity, readability, and additional information about the commands. A comment starts with the percent symbol (%) and can be written on a command line or on a separate line.

```
>> area_circle = 23.85      % square meters
```

MATLAB and Octave ignore comments so these have no effect on the execution of the program. Relevant comments provide important information that is not in the command lines.

```
>> y = 0.0       % position from the origin in meters
>> v = 0.0       % velocity in meters / second
>> a = -9.8      % acceleration in meters / second^2
```

3.9 Summary

The basic principles of programming using MATLAB and Octave are discussed. An introduction to the MATLAB and Octave commands is presented. Variables, constants, assignments, and built-in functions are discussed. Additional concepts such as comments for internal documentation and scientific notation are also discussed.

Key Terms

variables	constants	predefined constants
MATLAB/Octave prompt	assignment	assignment operator
mathematical expression	evaluation	built-in functions
scientific notation	programming	comments

Exercises

Exercise 3.1 Estimate the height of a building, given the height of a person, h, the distance from the building, D, and the elevation angle, θ, in degrees. Use a MATLAB/Octave command to compute the height of a building using the mathematical formula:

$$bh = h + D \times \tan(\theta \pi / 180)$$

Exercise 3.2 Use a MATLAB/Octave command to compute the distance between two points in the x-y plane using the mathematical expression:

$$d = \sqrt{(x_2-x_1)^2+(y_2-y_1)^2}$$

Exercise 3.3 Using arbitrary values for the variables, compute the value of q using a MATLAB/Octave assignment command and the following expression:

$$q = \frac{(u+v)^3}{\sqrt{x^2+y^2}}$$

Exercise 3.4 MATLAB and Octave provide function *fact()* to compute the factorial of a number. In mathematical notation, $n!$ is the factorial of n. Using arbitrary values for the variables, use a MATLAB/Octave assignment command to compute the following expression:

$$\frac{n!}{i!(n-i)!}$$

Exercise 3.5 Write a MATLAB expression that evaluates the following mathematical expression. Use arbitrary values for the variables mu, sigma and x.

$$\frac{e^{-\left(\frac{x-\mu}{\sigma\sqrt{2}}\right)^2}}{\sigma\sqrt{2\pi}}$$

Part II
Computational Models

Chapter 4

Introduction to Computational Models

4.1 Introduction

This chapter introduces the basic concepts and principles of computational modeling. These are: computational science, computational thinking, computational models, problem solving, and abstraction. An extremely simple problem is introduced and solved: the temperature conversion problem. A brief overview of selected MATLAB and Octave commands is included in order to explain the implementation of the solution to the problem discussed.

4.2 Preliminary Concepts

A **system** is part of the real world under study and that can be identified from the rest of its environment for a specific purpose. Such a system is called the real system because it is physically (or conceptually) part of the real world. A typical real-world system has structure and behavior. A system is composed of a set of entities (or components) that interact among themselves and with the environment to accomplish the system's goal. This interaction determines the behavior of the system.

A **model** is a representation of a real system or part of it. **Modeling** is the activity of building models. The model is simpler than the real system it represents, but it should be equivalent to the real system in all relevant aspects. In this sense, a model is an abstract representation of a real system.

Every model has a specific purpose and goal. A model only includes the aspects of the real system that were decided as being important, according to the initial requirements of the model. This implies that the limitations of the model have to be clearly understood and documented.

A **computational model** is a mathematical model implemented in a computer system and that usually requires high performance computational resources to execute. The model is used to study the behavior of a complex real system.

A computational model is often an implementation of a numerical solution to a complex nonlinear mathematical model that represents the real system and for which simple, intuitive analytical solutions are not readily available.

Abstraction is recognized as a fundamental and essential principle in problem solving and software development. This concept is extremely important in dealing with complex systems. Abstraction is the activity of hiding the details and exposing only the essential features of a particular system. Proper abstraction will result in a good model that helps solve the problem. Random addition and removal of features constitute another strategy for modeling, but usually does not result in a good model to solve the given problem. In modeling, one of the critical tasks is representing the various aspects of a system at different levels of abstraction. A good abstraction captures the essential elements of a system, purposely leaving out the rest. Abstraction is very useful in modeling large and complex systems, such as operating systems and real-time systems.

Computational thinking is the ability of a person to describe the requirements of a problem, design a mathematical solution (if possible) to the problem, and implement the solution on a computer. Computational thinking involves the following issues:

- Reasoning about computer problem solving

- The ability to describe the requirements of a problem and, if possible, design a mathematical solution that can be implemented on a computer

- The solution usually requires **multidisciplinary** and team approaches

- The solution normally leads to the construction of a **computational model**

Computer science is the study of computing structures and is based on mathematics. A primary modeling method is to use mathematical entities such as numbers, functions and sets to model problems and real-world systems. This is the *mathematical model* of the problem or system.

Computational science integrates concepts and principles from applied mathematics and computer science and applies them to the various scientific and engineering disciplines. Computational science is:

- An emerging multidisciplinary area

- The intersection of the more traditional sciences, engineering, applied mathematics, and computer science and focuses on the integration of knowledge for the development of problem-solving methodologies and tools that help advance the sciences and engineering areas. This is illustrated in Figure 4.1.

- An area that has as a general goal the development of high-performance computer models.

- An area that mostly involves multidisciplinary modeling and simulation.

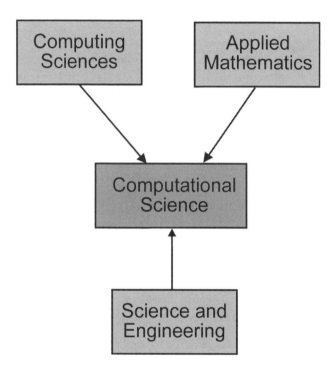

FIGURE 4.1: Computational science as an integration of several disciplines.

When a mathematical analytical solution of the model is not possible, a numerical and graphical solution is sought and experimentation with the model is carried out by changing the parameters of the model in the computer, and studying the differences in the outcome of the experiments. Further analysis and predictions of the operation of the model can be derived or deduced from these computational experiments.

Developing a computer solution to solve a problem entails the following general tasks:

1. Understand the problem description.

2. Design a mathematical representation of the problem, the mathematical model.

3. Find the mathematical solution to the model.

46 *Introduction to Elementary Computational Modeling*

4. Implement the mathematical solution to model of the problem.

5. Perform computations on the model; preferably the output should also be presented in visual form.

6. Interpret the results as solutions to the problem.

One of the goals of the general approach to problem solving is the focus on modeling the problem at hand, building or implementing the resulting solution using an appropriate tool environment (such as MATLAB or Octave) or with programming constructs in some appropriate programming language. The syntax and semantics of these constructs in the programming language are discussed elsewhere.

This book presents elementary material that links the various application areas of modeling with computer science. In this sense, the book concentrates on:

- General principles and concepts in modeling and problem solving

- Description and explanation of the process of developing and using computational models

4.3 A Simple Problem: Temperature Conversion

In this section, an extremely simple problem is introduced and solved: the temperature conversion problem. A basic sequence of steps is followed to solve this problem by developing a computational model implemented in MATLAB or Octave. A discussion of the MATLAB and Octave commands is included in order to explain the solution to the problem.

4.3.1 Initial Problem Statement

American tourists visiting Europe do not understand the units of temperature used in weather reports. The problem is to devise a computational model that indicates the temperature in Fahrenheit from a known temperature in Celsius.

4.3.2 Analysis and Conceptual Model

A brief analysis of the problem involves:

1. Understanding the problem. The main goal of the problem is to develop a temperature conversion facility from Celsius to Fahrenheit.

2. Finding the mathematical formula for the conversion of temperature from Celsius to Fahrenheit. Without this knowledge, we cannot derive a solution to this problem. The conversion formula is the mathematical model of the problem.

3. Knowledge of how to implement the mathematical model in a computer. We need to express the model in a particular computer tool or a programming language. The computer implementation must closely represent the model in order for it to be correct and useful.

4. Knowledge of how to test the program for correctness.

4.3.3 The Mathematical Model

The mathematical representation of the solution to the problem is the formula expressing a temperature measurement F in Fahrenheit in terms of the temperature measurement C in Celsius, which is:

$$F = 9/5 \times C + 32 \qquad (4.1)$$

Here C is a variable that represents the given temperature in degrees Celsius, and F is a derived variable, whose value depends on C.

A formal definition of a function is beyond the scope of this chapter. Informally, a *function* is a computation on elements in a set called the *domain* of the function, producing results that constitute a set called the *range* of the function. The elements in the domain are sometimes known as the input parameters. The elements in the range are called the output results.

Basically, a function defines a relationship between two (or more) variables, x and y. This relation is expressed as $y = f(x)$, so y is a function of x. For every value of x, there is a corresponding value of y. Variable x is the independent variable and y is the dependent variable.

The solution to the problem is the mathematical expression for the conversion of a temperature measurement in Celsius to the corresponding value in Fahrenheit. The mathematical formula expressing the conversion assigns a value to the desired temperature in the variable F, the dependent variable. The values of the variable C can change arbitrarily because it is the independent variable. The model uses real numbers to represent the temperature readings in various temperature units.

4.4 Using MATLAB and Octave

MATLAB is started by double-clicking on the MATLAB (or Octave) icon or invoking the application from the Start menu of Windows. The main MATLAB window, called the MATLAB Desktop, will then pop up and include the command window. Use the command window to type in all the MATLAB commands.

The commands are entered at the command prompt for MATLAB and Octave to execute; after, MATLAB and Octave execute the command and print out the result. Another command prompt is printed and MATLAB or Octave waits for another command. In this way, the user can interactively enter as many commands as needed.

4.4.1 Basic MATLAB and Octave Commands

Most of the simple commands in MATLAB and Octave are used with variables, constants, and arithmetic operators. Variables are named locations in memory where numbers, strings and other elements of data may be stored while the program is working. Variable names are combinations of letters and digits, but must start with a latter. To assign a value to a variable, use the assignment statement. This takes the form *variable = expression*. For example, the following MATLAB commands assign the 6.6 to variable *x* and uses an arithmetic expression to compute the value to assign to variable *y*. Figure 4.2 shows the same commands in an Octave console window.

```
>> x = 6.6
x =
    6.6
>> y = x * 2.4 + 36.0
y =
    51.84
```

4.4.2 The Computational Model

For the temperature conversion problem, the next step is to implement in MATLAB and Octave, the mathematical model represented by the equation, $F = 9/5 \times C + 32$. This model has a dependent variable, F, that corresponds to the value of the temperature in Fahrenheit, and an independent variable, C, for the given value of the temperature in Celsius. The mathematical expression (formula) allows the computation of F for arbitrary values given for C.

At the MATLAB or Octave prompt, the following command will compute

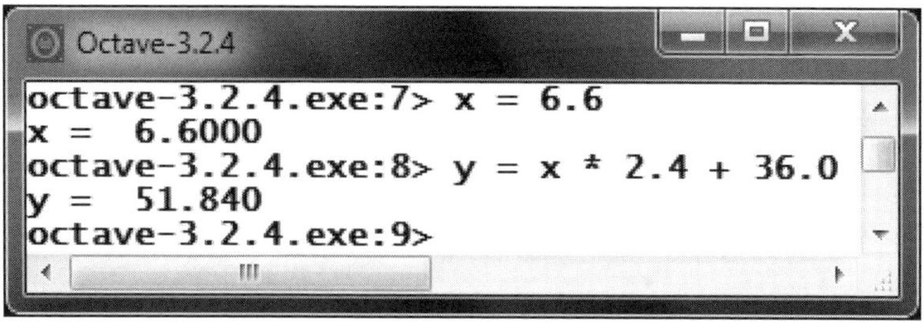

Introduction to Computational Models 49

```
Octave-3.2.4
octave-3.2.4.exe:7> x = 6.6
x =    6.6000
octave-3.2.4.exe:8> y = x * 2.4 + 36.0
y =    51.840
octave-3.2.4.exe:9>
```

FIGURE 4.2: Simple commands in an Octave window.

the Fahrenheit temperature given the value 5.0 for temperature in Celsius. The result is 41 degrees Fahrenheit computed by MATLAB or Octave; this value is assigned to variable F.

```
>> F = (9.0 / 5.0) * 5.0 + 32.0
F =
    41
```

The following commands compute the Fahrenheit temperature starting with a given value of 10.0 for the temperature in Celsius and then is repeated in increments of 5.0 degrees Celsius. The last computation is for a given value of 45.0 degrees Celsius.

```
>> F = (9.0 / 5.0) * 10.0 + 32.0
F =
    50
>> F = (9.0 / 5.0) * 15.0 + 32.0
F =
    59
>> F = (9.0 / 5.0) * 20.0 + 32.0
F =
    68
>> F = (9.0 / 5.0) * 25.0 + 32.0
F =
    77
>> F = (9.0 / 5.0) * 30.0 + 32.0
F =
    86
>> F = (9.0 / 5.0) * 35.0 + 32.0
F =
    95
>> F = (9.0 / 5.0) * 40.0 + 32.0
F =
    104
>> F = (9.0 / 5.0) * 45.0 + 32.0
F =
    113
```

Table 4.1 shows all the values of temperature in Celsius used to compute the corresponding temperature in Fahrenheit. This is a simple and short set of results of the original problem.

TABLE 4.1: Celsius and Fahrenheit temperatures.

Celsius	5	10	15	20	25	30	35	40	45
Fahrenheit	41	50	59	68	77	86	95	104	113

4.4.3 Using Data Lists with MATLAB and Octave

A convenient and more useful way to organize the data is to use a collection or an ordered list of values of the same variable. This type of list is known as an **array** or just as a **vector**, and basically consists of an ordered list of data values. In MATLAB and Octave, a vector can be represented as a row-vector or a column-vector.

For example, assume the temperature data are given for each hour of the day (a vector with 24 elements), and we need to plot the data against time. The vector with the time data can be created with the actual values to be used. The following command creates a vector called *time* with all the values of time to use.

```
>> time =
    [0, 100, 200, 300, 400, ..., 1900, 2000, 2100, 2200, 2300]
```

A much more convenient method of creating a vector is with the *colon* operator. This operator allows you to create an incremental vector of regularly spaced points by specifying:

```
lower_value:increment:upper_value
```

The following command creates a vector, *time*, with an initial value of 0, an increment of 100, and a final value of 2300.

```
octave-3.2.4.exe:17> time = 0:100:2300
time =
 Columns 1 through 11:
       0    100    200    300    400    500    600    700    800
     900   1000
 Columns 12 through 22:
    1100   1200   1300   1400   1500   1600   1700   1800   1900
    2000   2100
 Columns 23 and 24:
    2200   2300
```

A decrement may also be specified. The following command in MATLAB and Octave creates a time vector, *timed*, with values from 2300 to zero with decrements of 100.

```
octave-3.2.4.exe:18> time = 2300:-100:0
time =
 Columns 1 through 11:
    2300   2200   2100   2000   1900   1800   1700   1600   1500
    1400   1300
 Columns 12 through 22:
    1200   1100   1000    900    800    700    600    500    400
     300    200
 Columns 23 and 24:
     100      0
```

Once a vector is created, individual values of the data can be accessed using indexing. In a row vector, the left-most value has the index of one. The following commands create two vectors: *mvect1* and *mvect2*.

```
octave-3.2.4.exe:19> mvect1 = [1 5 7]
mvect1 =
    1    5    7
octave-3.2.4.exe:20> mvect2 = [8 10 24]
mvect2 =
    8   10   24
```

Each individual value in a vector is known as an element. An element of the array is accessed using an integer known as an index. The first element of vector *mvect1* is accessed by writing *mvect1(1)*. The following commands access several elements of vectors *mvect1* and *mvect2*.

```
octave-3.2.4.exe:21> mvect1(1)
ans =  1
octave-3.2.4.exe:22> mvect1(3)
ans =  7
octave-3.2.4.exe:23> mvect2(1)
ans =  8
octave-3.2.4.exe:24> mvect2(3)
ans =  24
```

With MATLAB and Octave, operators can be used with scalars (numbers) and vectors. For example, the following commands create vector *A* and multiply every element of vector *A* by 2. The resulting values are stored in vector *B*.

```
octave-3.2.4.exe:25> A = [1 2 3 4 5 6]
A =
    1    2    3    4    5    6
octave-3.2.4.exe:26> B = A .* 2
B =
    2    4    6    8   10   12
```

In a similar manner, other operators can be used with a vector and scalar. The following commands illustrate this.

Introduction to Computational Models 53

```
octave-3.2.4.exe:27> c = A .^ 2
c =
    1    4    9   16   25   36
octave-3.2.4.exe:28> d = A + 2
d =
    3    4    5    6    7    8
octave-3.2.4.exe:29> e = A - 2
e =
   -1    0    1    2    3    4
```

4.4.4 Implementation of Model with Data Lists

An improved solution of the temperature conversion problem involves using arrays. Using MATLAB or Octave, a vector C is created with a starting value of 0.0, an increment of 2.0, and a final value of 100.0.

```
octave-3.2.4.exe:35> C = 0.0:2.0:100.0
C =
 Columns 1 through 10
    0    2    4    6    8   10   12   14   16   18
 Columns 11 through 20
   20   22   24   26   28   30   32   34   36   38
 Columns 21 through 30
   40   42   44   46   48   50   52   54   56   58
 Columns 31 through 40
   60   62   64   66   68   70   72   74   76   78
 Columns 41 through 50
   80   82   84   86   88   90   92   94   96   98
 Column  51
  100
```

Having constructed vector C with the temperature values in degrees Celsius, the conversion formula can be applied to every value in vector C and results in the creation of vector F.

```
octave-3.2.4.exe:36> F = (9.0/5.0) * C + 32.0
F =
 Columns 1 through 6
   32.0000   35.6000   39.2000   42.8000   46.4000   50.0000
 Columns 7 through 12
   53.6000   57.2000   60.8000   64.4000   68.0000   71.6000
 Columns 13 through 18
   75.2000   78.8000   82.4000   86.0000   89.6000   93.2000
 Columns 19 through 24
   96.8000  100.4000  104.0000  107.6000  111.2000  114.8000
```

```
Columns 25 through 30
118.4000   122.0000   125.6000   129.2000   132.8000   136.4000
Columns 31 through 36
140.0000   143.6000   147.2000   150.8000   154.4000   158.0000
Columns 37 through 42
161.6000   165.2000   168.8000   172.4000   176.0000   179.6000
Columns 43 through 48
183.2000   186.8000   190.4000   194.0000   197.6000   201.2000
Columns 49 through 51
204.8000   208.4000   212.0000
```

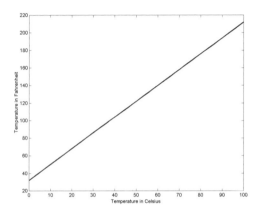

FIGURE 4.3: Plot of the temperature conversion.

Figure 4.3 shows the graph of the values of C and F. Note that the graph shows basically a straight line, which illustrates the linear dependency of F with respect to C. As also shown in Equation 4.1, there is a linear relationship between F and C, and the variable F is directly proportional to variable C.

The temperature values in Celsius are stored in vector C and the corresponding temperature values in Fahrenheit are stored in vector F. In computational modeling, it is often useful to get a visual representation of the mathematical relationships of the problem solution.

Listing 4.1 shows the complete sequence of MATLAB/Octave commands that solve the temperature conversion problem. These commands are stored in file mtconv.m.

Listing 4.1: Command file for computing temperature conversion.

```
1  % MATLAB/Octave script for Temperature conversion
2  % J. M. Garrido, October 2010
3  % File: mtconv.m
4  % C temperature in Celsius
5  % F temperature in Fahrenheit
6  % create an array from 0 to 100 degrees Celcius
7  C = 0.0:2.0:100.0
8  F = (9.0/5.0) * C + 32.0   % compute the temperature Fah
9  myp = plot (C,F)            % plot a graph
10 ylabel('Temperature in Fahrenheit')
11 xlabel('Temperature in Celsius')
12 set(myp, 'LineWidth', 2)
13 print -dpng tempconv.png
14 print -dtiff tempconv.tif
15 print -deps tempconv.eps
16 print -deps2 tempconv2.eps
17 print -dpdf tempconv.pdf
```

The following commands appear in lines 9 to 11 and plot the graph with the values of the temperature in Celsius on the X-axis and the values of temperature in Fahrenheit on the Y-axis.

```
plot (C,F)
box off
ylabel('Temperature in Fahrenheit')
xlabel('Temperature in Celsius')
```

4.5 Summary

The concepts and principles of computational modeling are discussed. These are: computational science, computational thinking, computational models, problem solving, and abstraction. An introduction to the MATLAB and Octave commands is presented in order to explain the solution to the problem discussed. The temperature conversion problem is discussed.

Key Terms		
modeling	computational thinking	models
problem solving	abstraction	computer science
mathematical model	pseudo-code	programming
programming language	implementation	MATLAB
Octave	problem solution	implementation

Exercises

Exercise 4.1 European tourists visiting the United States do not understand the units of temperature used in weather reports. The problem is to devise some mechanism for indicating the temperature in Celsius from a known temperature in Fahrenheit. Develop a computational model using MATLAB or Octave to solve this problem.

Exercise 4.2 Every student needs to know his or her average score from four tests already taken and for which the scores are already available. The problem is to determine the average score from the four test scores. Develop a computational model using MATLAB or Octave.

Exercise 4.3 Several lots of land are for sale. All lots are triangular with the length of the sides known in feet. The selling price of each lot is $12.55 per square foot. Develop a computational model using MATLAB or Octave that calculates the total selling price for each lot.

Exercise 4.4 Develop a similar model to the one in Exercise 4.1 that converts miles to kilometers, using MATLAB or Octave.

Exercise 4.5 Owners of SUVs need to know the miles per gallon every month. Given the amount of money spent to fill the gasoline tank, the number of miles driven since the last time the tank was filled, and the price per gallon, develop a computational model that computes the miles per gallon, using MATLAB or Octave.

Chapter 5

Computational Models and Simulation

5.1 Introduction

This chapter presents a high-level view of the computational model development process. Defining the problem statement and its specification, building the conceptual model, the mathematical model, the computational model, verification and validation. Other subtopics include simulation and simulation models; problem and model decomposition; software tools for implanting (programming), running the model, and the visualization of results. Types of models are briefly explained: continuous, discrete, deterministic, and stochastic models. Another simple problem is introduced and solved: the free-falling object. MATLAB and Octave commands are included in order to explain the solution to the problem discussed.

5.2 Categories of Computational Models

A system is composed of a set of *entities* (or components) that interact among themselves and with the *environment* to accomplish the system's goal. This interaction determines the behavior of the system. A system maintains its existence through the interaction of its components or parts. As discussed in the previous chapter, a model is a simplified representation of the actual system intended to promote understanding.

From the perspective of how the model changes state in time, computational models can be divided into two general categories:

1. Continuous models

2. Discrete-event models

A continuous model is one in which the changes of state in the model occur continuously with time. Often the state variables in the model are represented as continuous functions of time.

58   *Introduction to Elementary Computational Modeling*

For example, a model that represents the temperature in a boiler as part of a power plant can be considered a continuous model because the state variable that represents the temperature of the boiler is implemented as a continuous function of time. These types of models are usually modeled as a set of differential equations.

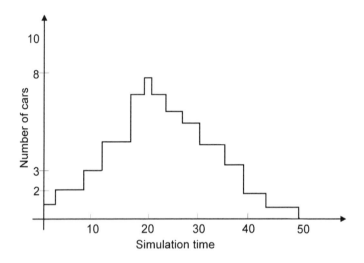

FIGURE 5.1: Discrete changes of number of cars in the queue.

In scientific and engineering practice, a simulation model of a real physical system is often executed through continuous simulation. These simulations of formal mathematical models are often carried out with software tools such as MATLAB and Simulink, which are computer programs designed for numeric computations and visualization.

A discrete-event model is one representing a system that changes its states at discrete points in time, i.e., at specific instants. The model of a simple car-wash system is a discrete-event model because an arrival event occurs, and causes a change in the state variable that represents the number of cars in the queue that are waiting to receive service from the machine (the server). This state variable and any other only changes its values when an event occurs, i.e., at discrete instants. Figure 5.1 illustrates the changes in the number of cars in the queue of the model for the simple car-wash system.

Depending on the variability of some parameters with respect to time, computational models can be separated into two categories:

1. Deterministic models

2. Stochastic models.

A deterministic model displays a completely predictable behavior. A stochastic model includes some uncertainty implemented with random variables, whose values follow a probabilistic distribution. In practice, a significant number of models are stochastic because the real systems being modeled usually include inherent uncertainty properties.

An example of a deterministic simulation model is a model of a simple car-wash system. In this model, cars arrive at exact specified instants (but at the same instants), and all have exact specified service periods (wash periods); the behavior of the model can be completely and exactly determined.

The simple car-wash system with varying car arrivals, varying service demand from each car, is a stochastic system. In a model for this system, only the averages of these parameters are specified together with a probability distribution for the variability of these parameters. Uncertainty is included in this model because these parameter values cannot be exactly determined.

5.3 Development of Computational Models

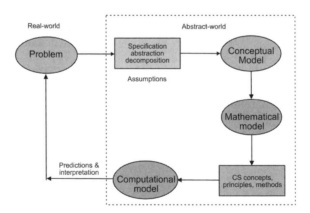

FIGURE 5.2: Development of computational models.

The process of developing computational models consists of a sequence of activities or stages that starts with the definition of modeling goals and is carried out in a possibly iterative manner. Because models are simplifications of reality there is a trade-off as to what level of detail is included in the model. If too little detail is included in the model one runs the risk of missing relevant interactions and the resultant model does not promote understanding. If too much detail is included in the model it may become overly complicated and actually preclude the development of understanding.

Computational models are generally developed in an iterative manner. After the first version of the model is developed, the model is executed, results from the execution run are studied, the model is revised, and more iterations are carried out until an adequate level of understanding is developed. The development process for a model involves the following general steps:

1. Definition of the **problem statement** for the computational model. This statement must provide the description of the purpose for building the model, the questions it must help to answer, and the type of expected results relevant to these questions.

2. Definition of the **model specification** to help define the conceptual model of the problem to be solved. This is a description of what is to be accomplished with the computational model to be constructed; and the assumptions (constraints), and domain laws to be followed. Ideally, the model specification should be clear, precise, complete, concise, and understandable. This description includes the list of relevant components, the interactions among the components, the relationships among the components, and the dynamic behavior of the model.

3. Definition of the **mathematical model**. This stage involves deriving a representation of the problem solution using mathematical entities and expressions and the details of the algorithms for the relationships and dynamic behavior of the model.

4. **Model implementation**. The implementation of the model can be carried out with a software environment such as MATLAB and Octave, in a simulation language, or in a general-purpose high-level programming language, such as Ada, C++, or Java. The simulation software to use is also an important practical decision. The main tasks in this phase are the coding, debugging, and testing of the software model.

5. **Verification** of the model. From different runs of the implementation of the model (or the model program), this stage compares the output results with those that would have been produced by a correct implementation of the conceptual and mathematical models. This stage concentrates on attempting to document and prove the correctness of the model implementation.

6. **Validation** of the model. This stage compares the outputs of the verified model with the outputs of a real system (or a similar already developed model). This stage compares the model data and properties with the available knowledge and data about the real system. Model validation

attempts to evaluate the extent to which the model promotes understanding.

A conceptual model can be considered as a high-level specification of the system and it is a descriptive model. It is usually described with some formal or semi-formal notation. For example, discrete-event simulation models are described with UML (the Unified Modeling Language) and/or extended simulation activity diagrams.

The conceptual model is formulated from the initial problem statement, informal user requirements, and data and knowledge gathered from analysis of previously developed models. The stages mentioned in the model development process are carried out at different levels of abstraction. Figure 5.3 illustrates the relationship between the various stages of model development and their abstraction level.

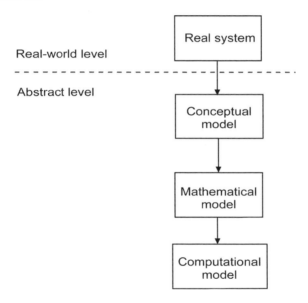

FIGURE 5.3: Model development and abstract levels.

5.4 Simulation: Basic Concepts

Simulation is a set of techniques, methods, and tools for developing a simulation model of a system and using and manipulating the simulation model to gain more knowledge about the dynamic behavior of a system. The purpose of simulation is to gain understanding about the behavior of the real system that the model represents.

A **simulation run** is the manipulation of a model in such a way that it operates on time or space to compress it, thus enabling one to perceive the interactions that would not otherwise be apparent because of their separation in time or space. A simulation run generally refers to a computerized execution of the model which is run over time to study the implications of the defined interactions.

An important goal of modeling and simulation is to gain a level of understanding of the interaction of the parts of a system, and of the system as a whole.

One of the great values of simulation is its ability to effect a time and space compression on the system, essentially allowing one to perceive, in a matter of minutes, interactions that would normally unfold over very lengthy time periods.

5.4.1 Simulation Models

A system is part of the real world under study and that can be separated from the rest of its environment for a specific purpose. Such a system is called the *real system* because it is inherently part of the real world.

Recall that a model is an abstract representation of a real system. The model is simpler than the real system, but it should be equivalent to the real system in all relevant aspects. The act of developing a model of a system is called *modeling*.

A simulation model is a computational model that has two main purposes:

1. To study some relevant aspects of the dynamic behavior of a system by observing the operation of the system, using the sequence of events or trace from the simulation runs

2. To estimate various performance measures

A simulation model is a computational model implemented as a set of procedures that when executed in a computer, *mimic* the behavior (in some relevant aspects) and the static structure of the real system. This type of model uses numerical methods as possibly the only way to achieve a solution. Simulation models include as much detail as necessary, that is, the representation of arbitrary complexity. The output of the model depends on its reaction to the following types of input:

- The passage of time;

- Data from the environment;

- Events (signals) from the environment.

The general purpose of a simulation model is to study the dynamic behavior of a system, i.e., the state changes of the model as time advances. The state of the model is defined by the values of its attributes, which are represented by state variables. For example, the number of waiting customers to be processed by a simple barbershop is represented as a state variable (an attribute), which changes its value with time. Whenever this attribute changes value, the system changes its state. For a model to be useful, it should allow the user to:

- Manipulate the model by supplying it with a corresponding set of inputs;

- Observe its behavior or output;

- Predict the behavior of the real system from the behavior of the model, under the same circumstances.

The reasons for developing a simulation model and carrying out simulation runs are:

- It may be too difficult, dangerous, and/or expensive to experiment with the real system.

- The real system is nonexisting; simulations are used to study the behavior of a future system (to be built).

As previously mentioned, the system behavior depends on the inputs from the environment. Figure 5.4 shows a simple model of a system interacting with its environment.

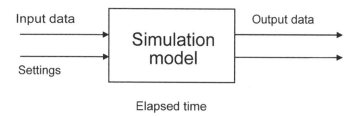

FIGURE 5.4: High-level view of a simulation model.

After a simulation model has been completely developed, the model is used to study the real system and solve the original problem. After being developed, a simulation model is used for:

- Designing experiments

- Performing simulation runs and collect data

- Analysis and interpretation - drawing inferences

- Documentation

5.4.2 Simulation Results

A simulation run is an experiment carried out on the simulation model for some period of observation; the time dimension is one of the most important in simulation. Several simulation runs are usually necessary in order to achieve some desired solution.

The results of experimenting with discrete-event simulation models, i.e., simulation runs, can be broken down into two sets of outputs:

1. Trace of all the events that occur during the simulation period and all the information about the state of the model at the instants of the events; this directly reflects the dynamic behavior of the model;

2. Performance measures — the summary results of the simulation run.

The trace allows the users to verify that the model is actually interacting in the manner according to the model's requirements. The performance measures are the outputs that are analyzed for estimates used for capacity planning or for improving the current real system.

5.5 Modular Decomposition

A problem is often too complex to deal with as a single unit. A general approach is to divide the problem into smaller problems that are easier to solve. The partitioning of a problem into smaller parts is known as *decomposition*. These small parts are known as *modules*, which are easier to manage.

Model design usually emphasizes modular structuring, also called modular decomposition. A problem is divided into smaller problems (or sub-problems) and a solution is designed for each sub-problem. Therefore, the solution to a problem consists of an assembly of the smaller solutions corresponding to each of the sub-problems. This approach is called modular design.

5.6 Average and Instantaneous Rate of Change

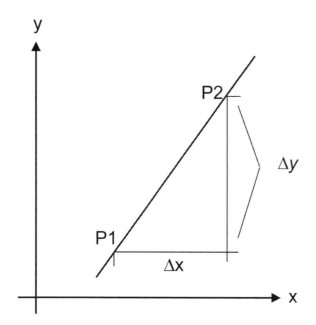

FIGURE 5.5: The slope of a line.

A mathematical function defines the relation between two (or more) variables. This relation is expressed as: $y = f(x)$. In this expression, variable y is a function of variable x, and x is the *independent variable* because for a given value of x, there is a corresponding value of y.

The average rate of change of a variable, y, with respect to a variable, x (the independent variable), is defined over a finite interval, Δx.

The Cartesian plane consists of two directed lines that perpendicularly intersect their respective zero points. The horizontal line is called the **x-axis** and the vertical line is called the **y-axis**. The point of intersection of the x-axis and the y-axis is called the **origin** and is denoted by the letter O.

The graphical interpretation of the average rate of change of a variable with respect to another is the **slope of a line** drawn in the Cartesian plane. The vertical axis is usually associated with the values of the dependent variable, y, and the horizontal axis is associated with the values of the independent variable, x.

Figure 5.5 shows a straight line on the Cartesian plane. Two points on the line, P_1 and P_2, are used to compute the slope of the line. Point P_1 is defined by two coordinate values (x_1, y_1) and point P_2 is defined by the coordinate values

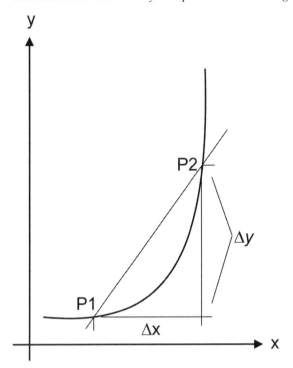

FIGURE 5.6: The slope of a secant.

(x_2, y_2). The horizontal distance between the two points, Δx, is computed by the difference $x_2 - x_1$. The vertical distance between the two points is denoted by Δy and is computed by the difference $y_2 - y_1$.

The *slope* of the line is the inclination of the line and is computed by the expression $\Delta y / \Delta x$, which is the same as the average rate of change of a variable y over an interval Δx. Note that the slope of the line is constant, on any pair of points on the line.

As mentioned previously, if the dependent variable y does not have a linear relationship with the variable x, then the graph that represents the relationship between y and x is a curve instead of a straight line. The average rate of change of a variable y with respect to variable x over an interval Δx, is computed between two points, P_1 and P_2. The line that connects these two points is called a **secant** of the curve. The average rate on that interval is defined as the slope of that secant. Figure 5.6 shows a secant to the curve at points P_1 and P_2.

The **instantaneous rate of change** of a variable, y, with respect to another variable, x, is the value of the rate of change of y at a particular value of x. This is computed as the slope of a line that is tangent to the curve at a point P.

Figure 5.7 shows a tangent of the curve at point P_1. The instantaneous rate

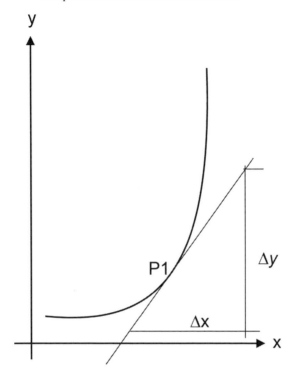

FIGURE 5.7: The slope of a tangent.

of change at a specified point P1 of a curve can be approximated by calculating the slope of a secant and using a very small interval; in different words, choosing Δx very small. This can be accomplished by selecting a second point on the curve closer and closer to point P1 (in Figure 5.7), until the secant almost becomes a tangent to the curve at point P1.

Examples of rate of change are: the average velocity, $\bar{v}$, computed by $\Delta y / \Delta t$, and the average acceleration $\bar{a}$, computed by $\Delta v / \Delta t$. These are defined over a finite time interval, Δt.

5.7 Area under a Curve

A curve is normally defined by a relationship between variables x and y, and this is expressed as a function, $y = f(x)$. A simple method for approximating the area under a curve between the bounds $x = a$ and $x = b$ is to partition the distance $(b-a)$ into several *trapezoids*, compute the areas of the trapezoids, and add all these areas.

A trapezoid is a four-sided region with two opposite sides parallel. In Figure 5.8, it is the two vertical sides that are parallel. The area of a trapezoid is the average length of the parallel sides, times the distance between them. The area of the trapezoid with width $\Delta x = x_2 - x_1$, is computed as the width, Δx, times the average height, $(y_2 + y_1)/2$.

$$A_t = \Delta x \frac{y_1 + y_2}{2}$$

For the interval of $[a,b]$ on variable x, it is divided into $n-1$ equal segments of length Δx. Any value of y_k is defined as $y_k = f(x_k)$. The trapezoid sum to compute the area under the curve for the interval of $[a,b]$ is defined by the summation of the areas of the individual trapezoids and is expressed as follows.

$$A = \sum_{k=2}^{k=n} [\Delta x \frac{1}{2}(f(x_{k-1}) + f(x_k))]$$

The larger the number of trapezoids, the better the approximation to the area under the curve. The area from a to b, with segments: $a = x_1 < x_2 < \ldots < x_n = b$ is given by:

$$A = \frac{b-a}{2n}[f(x_1) + 2f(x_2) + \ldots + 2f(x_{n-1}) + f(x_n)]$$

In MATLAB and Octave the area under a curve is approximated by the library function *trapz()*. The following command computes the area under a curve defined by variable y, with x as the independent variable, and using the trapezoid method. The command assigns the value of the area computed to variable A.

```
A = trapz(x, y)
```

5.8 The Free-Falling Object

A problem is solved by developing a computational model and executing it. The model development generally follows the computational model development process discussed previously. MATLAB and Octave are used to implement and run the model.

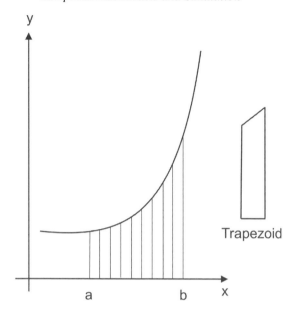

FIGURE 5.8: The area under a curve.

5.8.1 Initial Problem Statement

Students of physics need to know the vertical position and the velocity of a free-falling object as time passes. The solution to this problem is the calculation of vertical distance traveled and the velocity as the object approaches the ground. Several relevant questions related to the free-falling object need to be answered. Some of these are:

1. How does the acceleration of gravity affect the motion of the free-falling object?

2. How does the height of the free-falling object change with time, while the object is falling?

3. How does the velocity of the free-falling object change with time, while the object is falling?

4. How long does the free-falling object take to reach ground level, given the initial height, y_0? This question will not be answered here, it is left as an exercise.

5.8.2 Analysis and Conceptual Model

A brief analysis of the problem involves:

70   Introduction to Elementary Computational Modeling

1. Understanding the problem. The main goal of the problem is to develop a model to compute the vertical positions of the object from the point where it was released and the speed accordingly with changes in time.

2. Finding the relevant concepts and principles on the problem being studied. Studying the mathematical expressions for representing the vertical distance traveled and the vertical velocity of the falling object. This knowledge is essential for developing a mathematical model of the problem.

3. Listing the limitations and assumptions about the mathematical relationships found.

5.8.2.1   Assumptions

The main assumption for this problem is that near the surface of the earth, the acceleration due to the force of gravity is constant with value 9.8 m/s^2, which is also 32.15 ft/s^2. The second assumption is that the object is released from rest. The third important assumption is that the frictional drag due to resistance of the air is not considered.

5.8.2.2   Basic Definitions

The vertical motion of an object is defined in terms of displacement (y), velocity (v), acceleration (g), and time (t).

A time change, denoted by Δt, is a finite interval of time defined by the final time instance minus the initial time instance of the interval of time: $(t_2 - t_1)$. A change of displacement is denoted by Δy, and it represents the difference in the vertical positions of the object in a finite interval: $(y_2 - y_1)$. In a similar manner a change of velocity is denoted by Δv, and it represents the difference in the velocities in a finite interval: $(v_2 - v_1)$.

The velocity is the **rate of change** of displacement, and acceleration is the rate of change of velocity. The average velocity, denoted by $\bar{v}$, is the average rate of change of displacement with respect to time on the interval Δt. The average acceleration, denoted by $\bar{a}$, is the average rate of change of the velocity with respect to time on the interval Δt. These are defined by the following mathematical expressions:

$$\bar{v} = \frac{\Delta y}{\Delta t} \qquad \bar{a} = \frac{\Delta v}{\Delta t}$$

5.8.3   The Mathematical Model

The mathematical representation of the solution to the problem consists of the mathematical formulas expressing the the vertical displacement and the velocity of the object in terms of the time since the object was released and began free fall. Note that a general way to compute the average velocity, $\bar{v}$, is computed from the following expression:

$$\bar{v} = \frac{v_0 + v}{2},$$

with v_0 being the initial velocity and v the final velocity in that interval.

The mathematical model of the solution for a vertical motion of a free-falling object is considered next. Recall that in this model, the air resistance is ignored and the vertical acceleration is the constant $-g$. The vertical position as the object falls is expressed by the equation:

$$y = y_0 + v_0 t - \frac{gt^2}{2} \quad (5.1)$$

The velocity of the object at any time is given by the equation:

$$v_y = v_0 - gt, \quad (5.2)$$

where y is the vertical position of the object; t is the value of time; v is the vertical velocity of the object; v_0 is the initial vertical velocity of the object; and y_0 is the initial vertical position of the object. Equation 5.1 and Equation 5.2 represent the relationship among the variables: vertical position, vertical velocity, initial velocity, time instant, and initial vertical position of the object.

Note that in this model, the system state changes continuously with time. This type of state change is different compared with the example of the car-wash system. In the model of the car-wash system, the system state changes only at discrete instants of time. Another difference is that the model of the car-wash system includes an informal mathematical model, whereas the model of the free-falling object is a formal mathematical model, because it can be expressed completely by a set of mathematical equations (or expressions).

5.8.4 The Computational Model

For the free-falling object, the next step is to implement the mathematical model in MATLAB and Octave. The computational model has a dependent part that corresponds to the mathematical expression (formula) for the vertical position, y, and the vertical velocity, v_y, of the object and an independent part that allows arbitrary values given for time t. This really means that the computational model will use the equations (Equation 5.1 and Equation 5.2) defined previously.

5.8.4.1 Simple Implementation

For the MATLAB and Octave implementations of the mathematical model of the free-falling object, the constant parameters are first set. Note that the initial velocity, v_0, is zero. At the MATLAB prompt, the following commands will assign initial values to the parameter for the gravity acceleration, g in m/s^2, and the value of the initial height, y_0, in meters.

```
>> g = 9.8     % acceleration of gravity
g =
    9.8000
>> y0 = 40.0   % initial height in meters
y0 =
    40
```

The following MATLAB commands compute the vertical position (height) starting with a given value of y_0 for the initial height for time zero, $t = 0$. The calculations are then repeated for several values of time. Figure 5.9 shows the Octave console screen with the computations of the height of the free-falling object.

```
>> y = y0 - 0.5 * (g * 0.5 ^2) % height for t = 0.5
y =
    38.7750
>> y = y0 - 0.5 * (g * 0.7 ^2) % height for t = 0.7
y =
    37.5990
>> y = y0 - 0.5 * (g * 1.0 ^2) % height for t = 1
y =
    35.1000
>> y = y0 - 0.5 * (g * 1.2 ^2) % height for t = 1.2
y =
    32.9440
>> y = y0 - 0.5 * (g * 1.8 ^2) % height for t = 1.8
y =
    24.1240
>> y = y0 - 0.5 * (g * 2.0 ^2) % height for t = 2.0
y =
    20.4000
>> y = y0 - 0.5 * (g * 2.2 ^2) % height for t = 2.2
y =
    16.2840
>> y = y0 - 0.5 * (g * 2.5 ^2) % height for t = 2.5
y =
    9.3750
>> y = y0 - 0.5 * (g * 2.8 ^2) % height for t = 2.8
y =
    1.5840
>> y = y0 - 0.5 * (g * 3.0 ^2) % height for t = 3.0
```

y =
   -4.1000

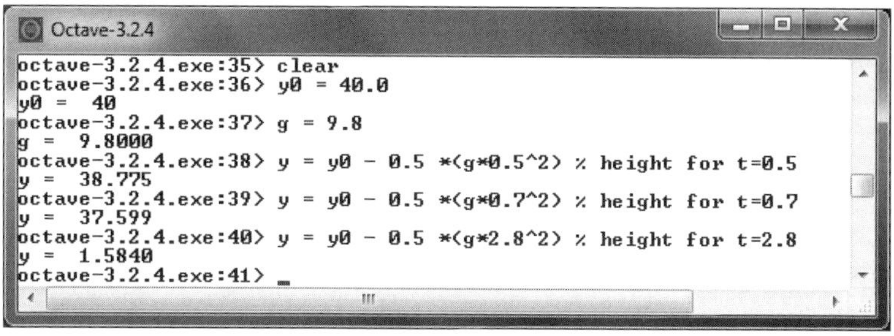

```
octave-3.2.4.exe:35> clear
octave-3.2.4.exe:36> y0 = 40.0
y0 = 40
octave-3.2.4.exe:37> g = 9.8
g = 9.8000
octave-3.2.4.exe:38> y = y0 - 0.5 *(g*0.5^2) % height for t=0.5
y = 38.775
octave-3.2.4.exe:39> y = y0 - 0.5 *(g*0.7^2) % height for t=0.7
y = 37.599
octave-3.2.4.exe:40> y = y0 - 0.5 *(g*2.8^2) % height for t=2.8
y = 1.5840
octave-3.2.4.exe:41>
```

FIGURE 5.9: Computing the height of the falling object in Octave.

The following MATLAB commands compute the values of the vertical velocity, v_y for the same values of time used to compute the values of height, y.

```
>> vy = - g * 0.5  % for t = 0.5
vy =
   -4.9000
>> vy = - g * 0.7  % for t = 0.7
vy =
   -6.8600
>> vy = - g * 1.0  % for t = 1.0
vy =
   -9.8000
>> vy = - g * 1.2  % for t = 1.2
vy =
   -11.7600
>> vy = - g * 1.8  % for t = 1.8
vy =
   -17.6400
>> vy = - g * 2.0  % for t = 2.0
vy =
   -19.6000
>> vy = - g * 2.2  % for t = 2.2
vy =
   -21.5600
>> vy = - g * 2.5  % for t = 2.5
vy =
   -24.5000
>> vy = - g * 2.8  % for t = 2.8
vy =
   -27.4400
>> vy = - g * 3.0  % for t = 3.0
vy =
   -29.4000
```

Table 5.1 shows most of the values used of the height and the vertical velocity computed with the values of time shown. This table represents a simple and short set of results of the original problem.

TABLE 5.1: Values of height and vertical velocity.

t	0.0	0.5	0.7	1.0	1.2	1.8	2.2	2.5	2.8
y	40.0	38.77	37.59	35.10	32.94	24.12	16.28	9.37	1.58
v_y	-0.0	-4.9	-6.86	-9.8	-11.7	-17.6	-21.5	-24.5	-27.4

5.8.4.2 Implementation with Arrays

Using MATLAB and Octave, vector *tf* is created with values of time starting value of 0.0, a final value of 3.5 seconds, and using 50 as the number of values in this vector. The MATLAB and Octave operation *linspace* is used to generate values of time starting at 0.0, ending at 3.5, using 50 different values. All the values of time used are stored in vector *tf*.

```
octave-3.2.4.exe:30> tf = linspace(0.0, 3.5, 50)
tf =
  Columns 1 through 7
      0      0.071   0.143   0.21    0.28    0.357   0.43
  Columns 8 through 14
      0.50   0.57    0.643   0.71    0.78    0.85    0.93
  Columns 15 through 21
      1.00   1.07    1.143   1.21    1.28    1.36    1.42
  Columns 22 through 28
      1.50   1.57    1.643   1.71    1.786   1.86    1.93
  Columns 29 through 35
      2.0    2.07    2.143   2.21    2.28    2.36    2.43
  Columns 36 through 42
      2.50   2.57    2.643   2.71    2.786   2.86    2.93
  Columns 43 through 49
      3.00   3.07    3.143   3.21    3.286   3.36    3.43
  Column 50
      3.5000
```

The following MATLAB command computes the vertical position (height) of the free-falling object for every value of time stored in vector *tf*. The values of height are stored in the new vector, *hf*.

```
octave-3.2.4.exe:31> hf = y0 - 0.5 * (g * tf.^2)
hf =
  Columns 1 through 7
     40.00   39.97   39.90   39.77   39.60   39.37   39.10
  Columns 8 through 14
     38.77   38.40   37.97   37.50   36.97   36.40   35.77
  Columns 15 through 21
     35.10   34.37   33.60   32.77   31.90   30.97   30.00
  Columns 22 through 28
     28.97   27.90   26.77   25.60   24.37   23.10   21.77
  Columns 29 through 35
     20.40   18.97   17.50   15.97   14.40   12.77   11.10
  Columns 36 through 42
      9.37    7.60    5.77    3.90    1.97   -0.00   -2.02
  Columns 43 through 49
```

```
 -4.10   -6.22   -8.40  -10.62  -12.90  -15.22  -17.60
Column 50
-20.0250
```

Listing 5.1 shows the complete sequence of MATLAB/Octave commands that compute the values of height in the problem of the free-falling object. These commands are stored in script file ffallobj.m.

Listing 5.1: MATLAB/Octave command file for computing the height.
```
1  % MATLAB/Octave command file
2  % Compute height of the free-falling object
3  % File: ffallobj.m
4  y0 = 40.0; % initail position
5  g = 9.8;
6  tf = linspace(0.0, 2.8, 50); % values of time
7  hf = y0 - 0.5 * (g * tf.^2); % computed values of height
8  ffobj = plot(tf, hf, 'k')
9  set(ffobj, 'LineWidth', 2)
10 box off
11 xlabel('Time')
12 ylabel('Height of free-falling object')
13 % store the plot in a figure
14 print -deps hffobj.eps
15 print -dpng hffobj.png
16 print -dtiff hffobj.tif
```

For a graphical illustration of the computations for the free-falling object, the next step is to plot the values of the two vectors, *tf* and *hf*, to get a visual representation of the problem solution. The following MATLAB commands plot a graph with the values of the time vector *tf* on the X-axis, and the values of the height vector *hf* on the Y-axis.

```
plot(tf, hf)
box off
xlabel('Time')
ylabel('Height of free-falling object')
```

Figure 5.10 shows the continuous change of the vertical position (height) of the free-falling object, with respect to time. Note that the graph represents a curve and not a straight line. This is because the mathematical relation between the variables y and t in Equation 5.1 is nonlinear; more specifically, it is a quadratic relationship between these two variables.

The following MATLAB and Octave command computes the vertical velocity of the free-falling object for every value of time stored in vector *tf*. The values of height are stored in the new vector, *vf*.

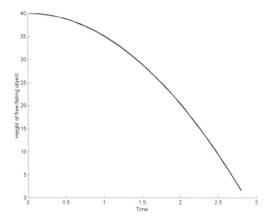

FIGURE 5.10: Plot of the values of height with time.

```
octave-3.2.4.exe:33> vf = -g * tf
vf =
  Columns 1 through 7
    0.00      -0.70     -1.40     -2.10     -2.80     -3.50     -4.20
  Columns 8 through 14
    -4.90     -5.60     -6.30     -7.00     -7.70     -8.40     -9.10
  Columns 15 through 21
    -9.80    -10.50    -11.20    -11.90    -12.60    -13.30    -14.00
  Columns 22 through 28
   -14.70    -15.40    -16.10    -16.80    -17.50    -18.20    -18.90
  Columns 29 through 35
   -19.60    -20.30    -21.00    -21.70    -22.40    -23.10    -23.80
  Columns 36 through 42
   -24.50    -25.20    -25.90    -26.60    -27.30    -28.00    -28.70
  Columns 43 through 49
   -29.40    -30.10    -30.80    -31.50    -32.20    -32.90    -33.60
  Column 50
   -34.3000
```

The next step is to plot the values of the two vectors, *tf* and *vf*, to get a visual representation of the problem solution. The following MATLAB and Octave commands plot a graph with the values of the time vector *tf* on the X-axis, and the values of the velocity vector *hf* on the Y-axis.

```
pvf = plot(tf, vf)
set(pvf, 'LineWidth', 1.5)
box off
xlabel('Time')
ylabel('Velocity of the falling object')
```

Listing 5.2 shows the complete sequence of MATLAB/Octave commands that compute the values of velocity in the problem of the free-falling object. These commands are stored in file `vffallobj.m`.

Listing 5.2: Command file for computing the velocity.

```
1  % MATLAB/Octave command file
2  % Velocity of the free-falling object
3  File: vffallobj.m
4  tf = linspace(0.0, 2.8, 50); % a vector for values of time
5  vf = -g * tf; % computed values of velocity
6  ffvobj = plot(tf, vf)
7  set(ffvobj, 'LineWidth', 2)
8  box off
9  xlabel('Time')
10 ylabel('Velocity of free-falling object')
11 % store the plot in a figure with eps, png, and tiff formats
12 print -deps vffobj.eps
13 print -dpng vffobj.png
14 print -dtiff vffobj.tif
```

Figure 5.11 shows the graph of the values of velocity with respect to time. Note that the graph is a straight line, which means that the relationship between velocity and time is a linear relation. This is consistent with Equation 5.2, which is a linear equation that relates the vertical velocity of the object and time.

The graph in Figure 5.11 also shows that the velocity is decreasing with time, starting with a value of 0 at time zero and decreasing linearly to almost $-35\ m/sec$ at time 3.5. The fact that the graph shows negative values of velocity is because the direction of the movement of the falling object is downward.

5.8.4.3 Computing the Rates of Change

The basic concepts of intervals and rate of change were discussed previously in Section 5.8.2.2. Δt is a **finite interval** of time defined by the final time instance minus the initial time instance $(t_f - t_i)$ of the interval. A change of displacement, Δy, is the difference $(y_f - y_i)$ in the vertical positions of the object over the time interval Δt, and a change of velocity, Δv, is the difference $(v_f - v_i)$ in the velocities in that time interval.

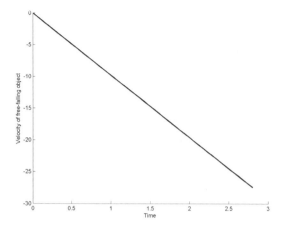

FIGURE 5.11: Plot of the values of velocity with time.

The average velocity, $\bar{v}$, is the rate of change of displacement with respect to time on the interval Δt, and is computed using $\Delta y/\Delta t$. The average acceleration, $\bar{a}$, is the rate of change of velocity with respect to time on the interval Δt, and is computed using $\Delta v/\Delta t$.

The rates of change of the displacement and of the velocity are computed from the data in Table 5.1. The first time interval considered is $[0, 0.5]$ seconds, Δt is 0.5 seconds. For this time interval, the vertical position y of the falling object changes from 40 meters to 38.77 meters, therefore Δy is 1.23 meters. The average rate of change of y, which is denoted by $\Delta y/\Delta t$ is 2.46 m/sec. In a similar manner, the vertical velocity changes from 0 to 4.9 m/sec for the same time interval. The values for the other intervals of time are computed in a similar manner and are shown in Table 5.2.

5.9 Summary

The computational model development process consists of a series of phases forc constructing a computational model. The various phases are: defining the problem statement and its specification, building the conceptual model, the mathematical model, the computational model, verification and validation. Simulation and simulation models are additional important concepts introduced in the study of computational models. Types of models are continuous,

TABLE 5.2: Values of the rates of change of height and vertical velocity of the free-falling object.

Time Interval	Δt	Δy	$\Delta y/\Delta t$	Δv	$\Delta v/\Delta t$
[0.0, 0.5]	0.5	1.23	2.46	4.9	9.8
[0.5, 0.7]	0.2	1.18	5.9	1.96	9.8
[0.0, 0.7]	0.7	2.41	3.44	6.86	9.8
[0.7, 1.0]	0.3	2.49	8.3	2.94	9.8
[1.0, 1.2]	0.2	2.16	10.8	1.96	9.8
[1.2, 1.8]	0.5	8.82	14.7	5.88	9.8
[1.8, 2.5]	0.7	14.76	21.08	6.86	9.8
[2.5, 2.8]	0.3	7.79	25.96	2.94	9.8

discrete, deterministic, and stochastic models. Computational models are implemented with standard programming languages and with tools such as MATLAB and Octave. Useful mathematical concepts discussed are: rates of change and the area under a curve.

Key Terms

conceptual model	mathematical model	computational model
model development	verification	validation
simulation	simulation model	decomposition
model decomposition	continuous model	discrete model
deterministic model	stochastic model	rate of change
average change	instantaneous change	area under curve

Exercises

Exercise 5.1 In the problem of the free-falling object, develop a computational model that computes the time the object takes to reach ground level and the vertical velocity at that time instance, given the initial height, y_0. Hint: derive a mathematical solution involving Equation 5.1 and Equation 5.2.

Exercise 5.2 Assume that the initial velocity, v_0, of the free-falling object is not zero. Develop a computational model that computes the time the object takes to reach ground level and the vertical velocity at that time instance, given the initial height, y_0. Hint: derive a mathematical solution involving Equation 5.1 and Equation 5.2.

Exercise 5.3 A large tent is to be set up for a special arts display in the main city park. The tent material is basically a circular canvas; the tent size can be adjusted by increasing or decreasing the distance from the center to the edge of the canvas, varying from 5 feet to 35 feet. The cost of the event will be paid based on the circular area occupied, $1.25 per square foot. Develop a computational model that takes several values of the distance from the center to the outer edge of the canvas (the radius) and provide the cost for the event.

Exercise 5.4 An account is set up that pays a guaranteed interest rate, compounded annually. The balance of the account will grow to a value at some point in the future that is known as the future value of the starting principal. Develop a computational model for calculating the future value given values of n years, P for the starting principal, and r for the rate of return expressed as a decimal (interest rate).

Chapter 6

Algorithms and Design Structures

6.1 Introduction

An algorithm is a detailed and precise sequence of steps that accomplishes a solution to a problem. An **algorithm** is designed to include the mathematical model and its solution. Because the purpose of computer problem solving is to design a solution to a problem, the algorithm then needs to be implemented in a computer and executed or run with the appropriate data for achieving the problem solution.

The implementation of an algorithm and the corresponding data descriptions is carried out with a software tool such as MATLAB, Octave, or with a standard programming language. This chapter contains the general concepts and explanations of design structures used in designing an algorithm and the high-level notations for describing it.

6.2 Problem Solving

Recall that the purpose of problem solving is to develop a computational model by implementing a computer solution to solve some real-world problem. The challenge is to find some method of solution or some way to approximate a solution to the problem. In order to design the solution to a problem, the first important step is to identify major parts of the problem:

- The given data

- Required results

- The necessary **transformations** to be carried out on the given data to produce the final desired results.

6.3 Algorithms

The algorithm is normally broken down into smaller tasks; the overall algorithm for a problem solution is decomposed into smaller algorithms each defined to solve a subtask.

An algorithm is a clear, detailed, precise, and complete description of the sequence of steps performed to produce the desired results. An algorithm can be considered the transformation on the given data and involves a sequence of commands or operations that are to be carried out on the data in order to produce the desired results. Figure 6.1 illustrates the basic notion of transformation on the data.

FIGURE 6.1: Transformation applied to the input data.

A computer implementation of an algorithm consists of a group of data descriptions and one or more sequences of instructions to the computer for producing correct results when given appropriate data. The implementation of an algorithm will normally be in the form of a program, which is written in an appropriate programming language and it tells the computer how to transform the given data into correct results.

An algorithm is normally described in a semiformal notation such as pseudo-code and flowcharts.

6.4 Describing Data

In a given problem, data consists of one or more data items. For every computation there is one or more data items, associated with some property in the problem, that are to be manipulated or transformed by the computations (computer operations). The input data is the set of data items that are transformed in order to produce the desired results.

To solve the problem, data descriptions together with algorithm descriptions are necessary. The algorithm is decomposed into the operations that manipulate the data and produce the results of the problem. The description for every data item is given by:

- The type of the data item
- A unique name to identify the data item
- An optional initial value

The name of a data item is an identifier and is arbitrarily chosen. The data type defines:

- The set of possible values that the data item may have
- The set of possible operations that can be applied to the data item

6.5 Notations for Describing Algorithms

Designing a solution to a problem involves design of an algorithm, which will be as general as possible in order to solve a family or group of similar problems. An algorithm can be described at several levels of abstraction. Starting from a very high-level and general level of description of a preliminary design, to a much lower-level that has more detailed description of the design.

To describe an algorithm, several notations are used; these are more informal and higher level than programming languages. An algorithmic notation is a set of general and informal rules used to describe an algorithm. Two widely used design notations are:

- Flowcharts
- Pseudo-code

6.5.1 Flowcharts

A flowchart is a visual representation of the flow of data and the operations on this data. A flowchart consists of a set of symbol blocks connected by arrows. The arrows that connect the blocks show the order for describing a sequence of design or action steps. The arrows also show the flow of data.

Several simple flowchart blocks are shown in Figure 6.2. Every flowchart block has a specific symbol. A flowchart always begins with a *start* symbol, which has an arrow pointing from it. A flowchart ends with a *stop* symbol, which has one arrow pointing to it.

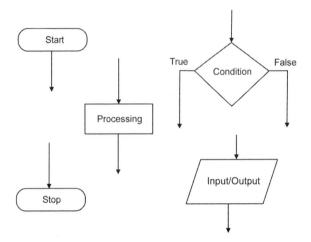

FIGURE 6.2: Simple flowchart symbols.

The *processing* or *transformation* symbol is the most common and general symbol, it is shown as a rectangular box. This symbol represents any computation or sequence of computations carried out on some data. There is one arrow pointing to it and one arrow pointing out from it.

The *selection* flowchart symbol has the shape of a vertical diamond, and it represents a *selection* of alternate paths in the sequence of design steps. This symbol is also shown in Figure 6.2. This symbol is known as a *decision block* or a conditional block because the sequence of instructions can take one of two directions in the flowchart.

The *input-output* flowchart symbol is used for a data input or output operation. There is one arrow pointing into the block and one arrow pointing out from the block.

An example of a simple flowchart with several basic symbols in shown in Figure 6.3. For larger or more complex algorithms, flowcharts are used mainly for the high-level description of the algorithms and pseudo-code for describing the details.

6.5.2 Pseudo-Code

Pseudo-code is a notation that uses a few simple rules and English for describing the algorithm that defines a problem solution. It can be used to describe relatively large and complex algorithms. It is relatively easy to convert the pseudo-code description of an algorithm to a computer implementation in a high-level programming language.

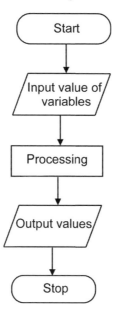

FIGURE 6.3: A simple flowchart example.

6.6 Algorithmic Structures

There are four fundamental design structures with which any algorithm can be described. These can be used with flowcharts and with pseudo-code notations. The basic design structures are:

- *Sequence*, any task can be broken down into a sequence of steps.

- *Selection*, this part of the algorithm takes a decision and selects one of several alternate paths of flow of actions. This structure is also known as alternation or conditional branch.

- *Repetition*, this part of the algorithm has a block or sequence of steps that are to be executed zero, one, or more times. This structure is also known as looping.

- *Input-output*, the values of variables are read from an input device (such as the keyword) or the values of the variables (results) are written to an output device (such as the screen)

6.6.1 Sequence

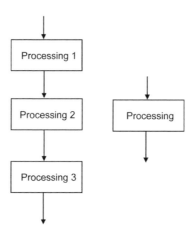

FIGURE 6.4: A flowchart with a sequence.

A sequence structure consists of a group of computational steps that are to be executed one after the other, in the specified order. The symbol for a sequence can be directly represented by a *processing* block of steps in a flowchart.

A sequence can also be shown as a flow of flowchart blocks. Figure 6.4 illustrates the sequence structure for both cases. The sequence structure is the most common and basic structure used in algorithmic design.

6.6.2 Selection

With the selection structure, one of several alternate paths will be followed based on the evaluation of a condition. Figure 6.5 illustrates the selection structure. In the figure, the actions or instructions in *Processing1* are executed when the condition is true. The instructions in *Processing2* are executed when the condition is false.

A flowchart example of the selection structure is shown in Figure 6.6. The condition of the selection structure is *len* > 0 and when this condition evaluates to true, the block with the action add 3 to k will execute. Otherwise, the block with the action decrement k is executed.

6.6.3 Repetition

The repetition structure indicates that a set of action steps are to be repeated several times. Figure 6.7 shows this structure. The execution of the actions in

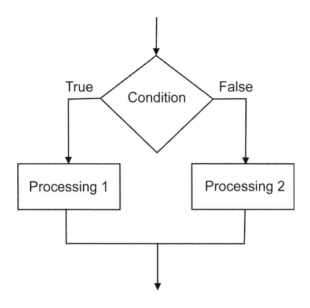

FIGURE 6.5: Selection structure in flowchart form.

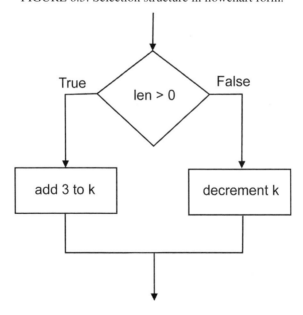

FIGURE 6.6: An example of the selection structure.

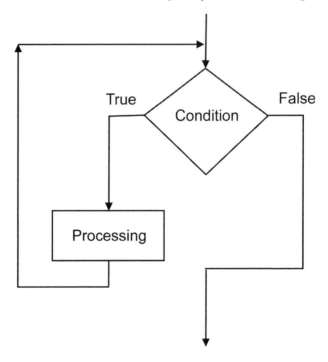

FIGURE 6.7: While loop of the repetition structure.

the *Processing* block are repeated while the condition is true. This structure is also known as the *while loop*.

A variation of the repetition structure is shown in Figure 6.8. The actions in the *Processing* block are repeated until the condition becomes true. This structure is also known as the *repeat-until* loop.

6.7 Implementation of Algorithms

As mentioned previously, detailed design of an algorithm is often first written in pseudo-code, which is a high-level descriptive notation. The software implementation of an algorithm is carried out writing the code in a suitable programming language and then executing the code.

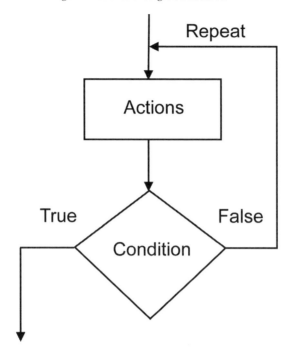

FIGURE 6.8: Repeat-until loop of the repetition structure.

6.7.1 Programming Languages

Programming languages, such as Ada, Fortran, Eiffel, C, C++, Java, MATLAB, and Octave, have well-defined syntax and semantic rules. The syntax is defined by a set of grammar rules and a vocabulary (a set of words). The legal sentences are constructed using sentences in the form of *statements*. There are two groups of words that are used to write the statements: *reserved words* and *identifiers*.

Reserved words are the keywords of the language and have a predefined purpose. These are used in most statements. Examples are: **for**, **end**, **function**, **while**, and **if**.

Identifiers are names for variables, constants, and functions that the programmer chooses, for example, *height*, *temperature*, *pressure*, *number_units*, and so on.

MATLAB and Octave, which are tools used for developing computational models include a simple and powerful programming language that is mainly adapted to mathematical applications.

This section discusses simple pseudo-code and computational model statements, such as the assignment and I/O statements.

6.7.2 Assignment and Arithmetic Expressions

The *assignment* statement is used to assign a value to a variable. This variable must be written on the left-hand side of the equal sign, which is the assignment operator. The value to assign can be directly copied from another variable, from a constant value, or from the value that results after evaluating an expression.

In the following example, the first statement assigns the constant value 35.5 to variable *length*. The second statement assigns the value 45 to variable *num*. The third assignment statement in the example, variable *z* is assigned the result of evaluating the expression, $y + 10.5 \times x^3$. In pseudo-code, these assignment statements are simply written as follows:

```
set length = 35.5
set num = 45
set z = y + 10.5x³
```

6.7.3 Simple Numeric Computations

An assignment is often written with an arithmetic expression that is evaluated and the result is assigned to a variable. For example, add 15 to the value of variable *x*, subtract the value of variable *y* from variable *z*. In pseudo-code, the **add** and **subtract** statements can be used in simple arithmetic expressions instead of the "+" and "−" operator symbols. The two statements are written as follows:

```
add 15 to x
subtract y from z
```

In the example, in the first statement, the new value of *x* is assigned by adding 15 to the previous value of *x*. In the second assignment, the new value of variable *z* is assigned the value that results after subtracting the value of *y* from the previous value of *z*.

To add or subtract 1 in pseudo-code, the **increment** and the **decrement** statements can be used. For example, the statement `increment j`, adds the constant 1 to the value of variable *j*. In a similar manner, the statement `decrement k` subtracts the constant value 1 from the value of variable *k*.

With MATLAB and Octave, the statements in the previous examples are written as:

```
x = x + 15
z = z - y
```

Algorithms and Design Structures 93

In the previous examples, only simple arithmetic expressions were used in the assignment statements. These are addition, subtraction, multiplication, and division. More complex calculations use various numerical functions in computational models, such as square root, exponentiation, and trigonometric functions. For example, the value of the expression $\cos p + q$ assigned to variable y and $\sqrt{x-y}$ assigned to variable q. The statements in pseudo-code are:

$$y = \cos(p) + q$$
$$q = \sqrt{x-y}$$

In MATLAB and Octave, these assignment statements use the mathematical functions *cos* and *sqrt*.

```
y = cos(p) + q
q = sqrt(x - y)
```

To assign the value of the mathematical expression $x^n \times y \times sin^{2m} x$ to variable z, the statement in pseudo-code is:

$$z = x^n \times y \times sin^{2m} x$$

In MATLAB and Octave, this assignment statement is written as follows:

```
z = x^n * y * (sin(x))^(2*m)
```

6.7.4 Simple Input/Output

Input and output statements are used to read (input) data values from the input device (e.g., the keyboard) and write (output) data values to an output device (mainly to the computer screen). The flowchart symbol is shown in Figure 6.9.

Using pseudo-code, the input statement reads a value of a variable from the input device (e.g., the keyboard). This statement is written with the keywords **read**, for input of a single data value and assign to a variable.

The following two lines of pseudo-code include the general form of the input statement and an example that uses the **read** statement to read a value of variable y.

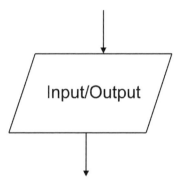

FIGURE 6.9: Flowchart data input/output symbol.

    read ⟨ var_name ⟩

    read y

The input statement implies an assignment statement for the variable y, because the variable changes its value to the new value that is read from the input device.

The MATLAB and Octave programming languages provide two basic functions for console input/output. The function for input reads the value of a variable from the input device. The name of the function is **input**, and is used with an assignment statement for input of a single data value. A string message is included to prompt the user for input of a data value. The general form for input in MATLAB and Octave is shown in the following line of code.

    ⟨ var_name ⟩ = **input** (⟨ str_message ⟩)

The previous example for reading the value of variable y is written in MATLAB and Octave as follows:

```
y = input('Enter value of altitude: ')
```

For output, pseudo-code has a display statement that can be used for the output of a list of variables and literals; it is written with the keyword **display**.

The output statement writes the value of one or more variables to the output device. The variables do not change their values. The general form of the output statement in pseudo-code is:

    display ⟨ data_list ⟩

display "value of x= ", x, "value of y = ", y

In MATLAB and Octave, function *moutput* included with the book software, accepts a single string and data value, so it can be used individually to display every variable and its value. The general form for this function call in MATLAB and Octave is:

moutput (⟨ *str_message* ⟩, ⟨ *var_name* ⟩)

The previous example outputs the value of two variables, x and y, so the output function is used two times.

```
moutput('value of x= ', x)
moutput('value of y = ', y)
```

6.8 Computing Area and Circumference

For this problem, a computational model is developed that computes the area and circumference of a circle. The input value of radius is read from the keyboard and the results written to the screen.

6.8.1 Specification

The specification of the problem can be described as a high-level algorithm in informal pseudo-code notation:

1. Read the value of the radius of a circle, from the input device.

2. Compute the area of the circle.

3. Compute the circumference of the circle.

4. Output or display the value of the area of the circle to the output device.

5. Output or display the value of the circumference of the circle to the output device.

6.8.2 Algorithm with the Mathematical Model

A detailed description of the algorithm and the corresponding mathematical model follows:

1. Read the value of the radius r of a circle, from the input device.
2. Establish the constant π with value 3.14159.
3. Compute the area of the circle, $area = \pi \times r^2$.
4. Compute the circumference of the circle $cir = 2 \times \pi \times r$.
5. Print or display the value of *area* of the circle to the output device.
6. Print or display the value of *cir* of the circle to the output device.

The following lines of pseudo-code completely define the algorithm.

>   **read** r
> $\pi = 3.1416$
> $area = \pi \times r^2$
> $cir = 2 \times \pi \times r$
>   **display** "Area = ", area, " Circumference = ", cir

The computational model is implemented in MATLAB and Octave and the code is stored in a script file named *areacir.m*, which is shown in Listing 6.1.

Listing 6.1: Commands for computing the area and circumference.
```
1  % MATLAB/Octave command file
2  % Compute the area and circumference of
3  % a circle, given its radius.
4  % File: areacir.m
5  % pi = 3.14159    % predefined constant
6  r = input ('enter value of radius: ')
7  % compute area of circle
8  area = pi * r^2
9  % compute circumference of circle
10 cir = 2.0 * pi * r
11 % now output the results
12 moutput('Area of circle is: ', area)
13 moutput('Circumference of circle is: ', cir)
```

The script file *areacir.m* is run by typing its name at the MATLAB or Octave prompt. The command at the prompt, input value used, and the results displayed are shown in the following listing.

```
octave-3.2.4.exec:33> areacir
enter value of radius: 3.15
r =      3.1500
area = 31.172
cir = 19.792
Area of circle is: 31.172
Circumference of circle is: 19.792
```

6.9 Summary

A computer program is the implementation of a computational model as the computer solution to a problem. The program is written in an appropriate programming language. The program includes the data descriptions and the algorithm. Data descriptions involve identifying, assigning a name, and assigning a type to every data item used in the problem solution. Data items are variables and constants.

The precise, detailed, and complete description of a solution to a problem is known as an algorithm. The notations to describe algorithms are flowcharts and pseudo-code. Flowcharts are a visual representation of the execution flow of the various instructions in the algorithm. Pseudo-code is an English-like notation to describe algorithms.

The design structures are sequence, selection, repetition, and input-output. These algorithmic structures are used to specify and describe any algorithm. The implementation of the algorithm is carried by programs written in a tools language such as MATLAB or Octave, or a conventional programming language such as C++, Java, Ada, Eiffel, and others.

Key Terms			
algorithm	flowcharts	pseudo-code	variables
constants	declarations	structure	sequence
action step	selection	repetition	input/output
statements	data type	identifier	design

Exercises

Exercise 6.1 Write the algorithm and data descriptions for computing the area of a triangle. Use flowcharts.

Exercise 6.2 Write the algorithm and data descriptions for computing the area of a triangle. Use pseudo-code.

Exercise 6.3 Develop a computational model for computing the area of a triangle.

Exercise 6.4 Write the algorithm and data descriptions for computing the perimeter of a triangle. Use pseudo-code.

Exercise 6.5 Write the algorithm and data descriptions for computing the perimeter of a triangle. Use flowcharts.

Exercise 6.6 Develop a computational model with a computational model program for computing the perimeter of a triangle.

Exercise 6.7 Write the data and algorithm descriptions in flowchart and in pseudo-code to compute the conversion from a temperature reading in degrees Fahrenheit to Centigrade. The algorithm should also compute the conversion from Centigrade to Fahrenheit.

Exercise 6.8 Develop a computational model with a computational model program to compute the conversion from a temperature reading in degrees Fahrenheit to Centigrade. The program should also compute the conversion from Centigrade to Fahrenheit.

Exercise 6.9 Write an algorithm and data descriptions in flowchart and pseudo-code to compute the conversion from inches to centimeters and from centimeters to inches.

Exercise 6.10 Develop a computational model with a computational model program to compute the conversion from inches to centimeters and from centimeters to inches.

Chapter 7

Selection

7.1 Introduction

As explained in the previous chapter, in order to completely describe algorithms, four design structures are used: sequence, selection, repetition, and input/output. This chapter explains the selection structure and the corresponding statements in pseudo-code, MATLAB, and Octave for implementing computational models.

Conditions are expressions that evaluate to a truth value (true or false). Conditions are used in the selection statements. Simple conditions are formed with relational operators for comparing two data items. Compound conditions are formed by joining two or more simple conditions with a *logical* operator.

The solution to a quadratic equation is discussed as an example of applying the selection statements.

7.2 Selection Structure

The selection design structure is also known as alternation, because alternate paths are considered based on the evaluation of a condition. This section covers describing the selection structure with flowcharts and pseudo-code, the concepts associated with conditional expressions, and implementation in MATLAB or Octave.

7.2.1 General Concepts of the Selection Structure

The selection structure is used for decision making in the logic of a program. Figure 7.1 shows the selection design structure using a flowchart. Two possible paths for the execution flow are shown. The condition is evaluated, and one of the paths is selected. If the condition is true, then the left path is selected and *Processing 1* is performed. If the condition is false, the other path is selected and *Processing 2* is performed.

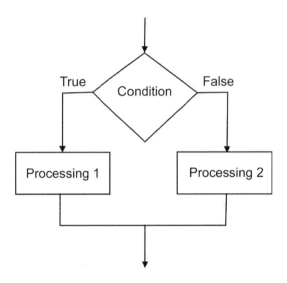

FIGURE 7.1: Flowchart of the selection structure.

7.2.2 Selection with Pseudo-Code

With pseudo-code, the selection structure is written with an **if** statement, also known as an if-then-else statement. This statement includes three sections: the condition, the then-section, and the else-section. The else-section is optional. Several keywords are used in this statement: **if**, **then**, **else**, and **endif**. The general form of the **if** statement in pseudo-code is:

```
if 〈 condition 〉
   then
       〈 statements in Processing 1 〉
   else
       〈 statements in Processing 2 〉
endif
```

When the condition is evaluated, only one of the two alternatives will be carried out: the one with the statements in *Processing 1* if the condition is true, or the one with the statements in *Processing 2* if the condition is not true. The pseudo-code of the selection structure in Figure 7.1 is written with an **if** statement as follows:

```
if condition is true
then
    perform instructions in
Processing 1
else
    perform instructions in
Processing 2
endif
```

7.2.3 Selection with MATLAB and Octave

Similarly to pseudo-code, MATLAB or Octave include the **if** statement that allows checking for a certain condition and executing statements if the condition is met. Note that MATLAB or Octave do not have the **then** keyword, and uses **end** instead of **endif**. The general form of the **if** statement in MATLAB or Octave is:

```
if ⟨ condition ⟩
    ⟨ statements in Processing 1 ⟩
else
    ⟨ statements in Processing 2 ⟩
end
```

The first line specifies the condition to check. If the condition evaluates to **true**, MATLAB executes the body of the first indented sequence of statements, otherwise the second indented sequence of statements between the **else** and the **end** are executed.

MATLAB and Octave do not require you to indent the body of an **if** statement, but it makes the code more readable.

7.2.4 Conditional Expressions

Conditional or Boolean expressions are used to form conditions. A condition consists of an expression that evaluates to a truth value, **true** or **false**.

A conditional expression can be constructed by comparing the values of two data items and using a relational operator. The following list of relational operators can be included in a conditional expression:

1. Equal, $=$

2. Not equal, $\neq$

3. Less than, $<$

4. Less than or equal to, $\leq$

5. Greater than, $>$

6. Greater than or equal to, $\geq$

With pseudo-code, these relational operators are used to construct conditions as in the following examples:

```
temperature ≤ 43.5
```
$x \geq y$
$a = b$

Pseudo-code syntax also provides additional keywords that can be used for the relational operators. For example, the previous conditional expressions can also be written as follows:

```
temperature  less or equal to  43.5
x   greater or equal to  y
a   equal to  b
```

With MATLAB or Octave, the relational operators are written using the following symbols: ==, for "equal," and ~=, for "not equal." Note that == is the operator that tests equality, and = is the assignment operator. If = is used in the condition of an **if** statement, it will result in a syntax error. The other relational operators in MATLAB or Octave are: <, >, <=, and >=.

In programming, a variable that contains a truth or logical value is often called a *flag* because it flags the status of some condition.

7.2.5 Example with Selection

In this example, the condition to be evaluated is: *len* > 0, and a decision is taken to select which computation to carry out on variable *k*. Figure 7.2 shows the flowchart for part of the algorithm that includes this selection structure. In pseudo-code, this example is written as:

```
if len  greater than  0
then
    add 3 to k
else
    decrement k
endif
```

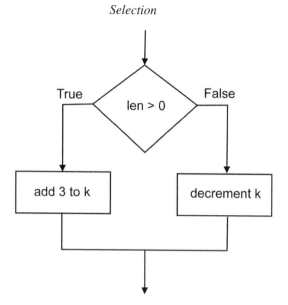

FIGURE 7.2: Example of selection structure.

The previous example in pseudo-code can also be written in the following manner:

if len >0
then
    k = k + 3
else
    k = k - 1
endif

In MATLAB or Octave, the previous example is written as follows:

```
if len > 0
    k= k+3
else
    k=k-1
end
```

7.3 Complex Numbers with MATLAB and Octave

Complex numbers have two parts: a **real** component and an **imaginary** component. These two numeric values are normally treated separately in arithmetic calculations. For example, consider a variable, w, assigned the complex value: $2 + 5i$. The real component of w is 2 and the imaginary component is 3. The special symbol i that is used with complex numbers has value $\sqrt{-1}$.

The relationship between the real and imaginary parts of a complex number can be visualized in a **complex plane**. A complex value is indicated as a point in this plane. The vertical axis is used to indicate the value of the imaginary component and the horizontal axis to indicate the value of the real component. This is the set of rectangular coordinates. Complex numbers can also be represented with polar coordinates, using two values: the magnitude and the angle in radians from the horizontal axis. Figure 7.3 illustrates the coordinates used in the complex plane for a point P that represents a complex number.

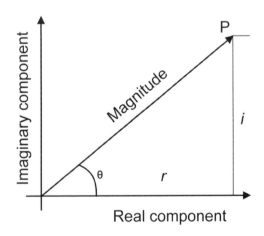

FIGURE 7.3: Complex number P in the complex plane.

In the following listing, the first three lines show sqrt(-1) entered at the MATLAB prompt and the answer printed. The next lines of MATLAB or Octave code show w = 2 + 5i entered at the prompt and the response from MATLAB.

```
>> sqrt(-1)
ans =
        0 + 1.0000i
>> w = 2 + 5i
w =
   2.0000 + 5.0000i
```

MATLAB and Octave provide complex arithmetic and several functions for use with complex numbers. The following lines of MATLAB or Octave code include an assignment of a complex number to variable x, the complex addition of variables w and y, and the result assigned to variable y. Note that since variables w and x are complex variables, variable y also becomes a complex variable.

```
>> x = 10.5-2.45i
x =
   10.5000 - 2.4500i
>> y = w+x
y =
   12.5000 + 2.5500i
```

Functions *real* and *imag* are used for accessing the real and imaginary parts of a complex number. For example, the following lines of MATLAB and Octave code get the value of the real and imaginary parts of complex variable y and assign these to variables *ry* and *iy*:

```
>> ry = real(y)
ry =
   12.5000
>> iy = imag(y)
iy =
    2.5500
```

In the following command lines, functions *abs* and *angle* are called in MATLAB and Octave to get the value of the magnitude of complex variable y and the angle from the horizontal axis in radians. These values are assigned to variables *magy* and *thetay*.

```
>> magy = abs(y)
magy =
   12.7574
>> thethay = angle(y)
thethay =
    0.2012
```

7.4 A Computational Model with Selection

The following problem involves developing a computational model that includes a quadratic equation, which is a simple mathematical model of a second-degree equation. The solution to the quadratic equation involves complex numbers.

7.4.1 Analysis and Mathematical Model

The goal of the solution to the problem is to compute the two roots of the equation. The mathematical model is defined in the general form of the quadratic equation (second-degree equation):

$$ax^2 + bx + c = 0$$

The given data for this problem are the values of the coefficients of the quadratic equation: a, b, and c. Because this mathematical model is a second-degree equation, the solution consists of the value of two roots: x_1 and x_2.

7.4.2 Algorithm for General Solution

The general solution gives the value of the two roots, when the value of the coefficient a is not zero ($a \neq 0$). The values of the two roots are:

$$x_1 = \frac{-b + \sqrt{b^2 - 4ac}}{2a} \quad x_2 = \frac{-b - \sqrt{b^2 - 4ac}}{2a}$$

The expression inside the square root, $b^2 - 4ac$, is known as the discriminant. If the discriminant is negative, the solution will involve complex roots. Figure 7.4 shows the flowchart for the general solution and the following listing is a high-level pseudo-code version of the algorithm.

```
Input the values of the coefficients
a,   b,   and   c
Calculate value of the discriminant
if the value of the discriminant is
less than zero
        then calculate the two complex
roots
        else calculate the two real roots
endif
display the value of the roots
```

Selection

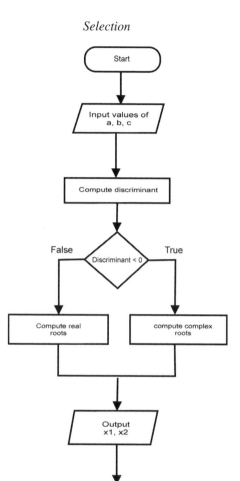

FIGURE 7.4: High-level flowchart for the quadratic equation.

7.4.3 Detailed Algorithm

The algorithm in pseudo-code notation for the solution of the quadratic equation is:

```
read the value of a from the input
device
read the value of b from the input
device
read the value of c from the input
device
compute the discriminant, $disc = b^2 - 4ac$
```
if discriminant less than zero
then
```
        // roots are complex
        compute $x1 = (-b + \sqrt{disc})/2a$
        compute $x2 = (-b - \sqrt{disc})/2a$
```
else
```
        // roots are real
        compute $x1 = (-b + \sqrt{disc})/2a$
        compute $x2 = (-b - \sqrt{disc})/2a$
```
endif
```
display values of the roots: x1 and
x2
```

Listing 7.1 shows the MATLAB/Octave implementation of the algorithm for the solution of the quadratic equation, which is stored in the script file solquadra.m.

Note the two functions that are used for displaying values. Function *disp* is predefined in MATLAB and Octave for displaying a value, function *moutput* is defined in the script file moutput.m is used for displaying a text string and the value of a variable.

Listing 7.1 MATLAB/Octave script for solving a quadratic equation.

```
1  % Matlab/Octave script to compute the roots of a
2  %    quadratic equation
3  % File: solquadra.m
4  % Assume value of coefficient a is not zero
5  a = input ('Enter value of coefficient a: ');
6  b = input ('Enter value of coefficient b: ');
7  c = input ('Enter value of coefficient c: ');
8  %
9  % compute value of the discriminant
10 disc=b^2 - 4.0*a*c;
11 moutput('Discriminant: ', disc)
12 if disc < 0
13    disp('Roots are complex')
14 else
15    disp('Roots are real')
16 end
```

```
17 % calculate roots
18 x1 = (-b + sqrt(disc))/(2.0*a);
19 x2 = (-b - sqrt(disc))/(2.0*a);
20 % now display the results
21 moutput('Value of root 1 is: ', x1)
22 moutput('Value of root 2 is: ', x2)
```

The following MATLAB and Octave command lines execute the `solquadra.m` script. It prompts the user for the three values of the coefficients, calculates the roots, then displays the value of the roots. Figure 7.5 shows the Octave console for solving the quadratic equation.

```
EDU solquadra
Enter value of coefficient a: 1.5
Enter value of coefficient b: 2.25
Enter value of coefficient c: 5.5
Discriminant: -27.9375
Roots are complex
Value of root 1 is: -0.75+1.7619i
Value of root 2 is: -0.75-1.7619i
```

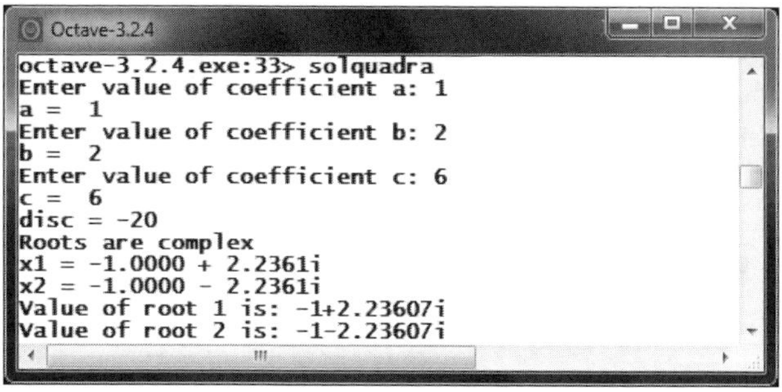

FIGURE 7.5: Solving the quadratic equation in Octave.

7.5 Multilevel Selection

There are often decisions that involve more than two alternatives. The general **if** statement with multiple paths is used to implement this structure. The case structure is a simplified version of the selection structure with multiple paths.

7.5.1 General Multipath Selection

The **elseif** clause is used to expand the number of alternatives. The **if** statement with n alternative paths has the general form:

```
if ⟨ condition ⟩
   then
        ⟨ block1 ⟩
   elseif ⟨ condition2 ⟩
   then
        ⟨ block2 ⟩
   elseif ⟨ condition3 ⟩
   then
        ⟨ block3 ⟩
   else
        ⟨ blockn ⟩
endif
```

Each block of statements is executed when that particular path of logic is selected. This selection depends on the conditions in the multiple-path **if** statement that are evaluated from top to bottom until one of the conditions evaluates to true. The following example shows the **if** statement with several paths.

```
if y > 15.50
then increment x
elseif y > 4.5
then add 7.85 to x
elseif y > 3.85
then compute x = y*3.25
elseif y > 2.98
then x = y + z*454.7
else x = y
endif
```

In MATLAB and Octave, this example is very similar as the following listing shows.

```
if y > 15.50
   x = x + 1
elseif y > 4.5
   x = x + 7.85
elseif y > 3.85
   x = y*3.25
elseif y > 2.98
   x = y + z*454.7
else
   x = y
end
```

7.5.2 The Case Structure

The case structure is a simplified version of the selection structure with multiple paths. In pseudo-code, the **case** statement evaluates the value of a single variable or simple expression of a number or a text string and selects the appropriate path. The case statement also supports a block of multiple statements instead of a single statement in one or more of the selection options. The general case structure is:

    case ⟨ select_var ⟩ **of**
      value   var_value1 : ⟨ block1 ⟩
      value   var_value2 : ⟨ block2 ⟩
        ...
      value   var_valuen : ⟨ blockn ⟩
    endcase

In the following example, the pressure status of a furnace is monitored and the pressure status is stored in variable *press_stat*. The following case statement first evaluates the pressure status and then assigns an appropriate text string to variable *msg*, then this variable is displayed.

        case press_stat **of**
                value 'D' : msg = "Very dangerous"
            value 'A' : msg = "Alert"
            value 'W' : msg = "Warning"
            value 'N' : msg = "Normal"
                value 'B' : msg = "Below normal"
        endcase
          ...
        display "Pressure status: ", msg

The default option of the **case** statement can also be used by writing the keywords **default** or **otherwise** in the last case of the selector variable. For example, the previous example can be enhanced by including the default option in the case statement:

```
case press_stat of
         value 'E' : msg = "Very dangerous"
     value 'D' : msg = "Alert"
     value 'H' : msg = "Warning"
     value 'N' : msg = "Normal"
         value 'B' : msg = "Below normal"
     otherwise
         msg = "Pressure rising"
endcase
    . . .
display "Pressure status: ", msg
```

In MATLAB and Octave, the **case** statement is called **case/switch**. The general form of this statement is:

```
switch 〈 select_var 〉
  case   var_value1   〈 block1 〉
  case   var_value2   〈 block2 〉
   . . .
  case   var_valuen   〈 blockn 〉
  otherwise
        〈 blockn 〉
end
```

In the previous example of the pressure status of a furnace in variable *press_stat*, the following MATLAB and Octave switch/case statement first evaluates the pressure status and then assigns an appropriate text string to variable *msg*, then this variable is displayed.

```
switch press_stat
   case 'D'
      msg = 'Very dangerous'
   case 'A'
      msg = 'Alert'
   case 'W'
      msg = 'Warning'
   case 'N'
      msg = "Normal"
   case 'B'
      msg = 'Below normal'
   otherwise
      msg = 'Pressure rising'
end
moutput('Pressure status: ', msg)
```

7.6 Complex Conditions

Logical operators **and**, **or**, and **not** are used to construct more complex conditional expressions from simpler conditions. The general form of a complex condition from two simple conditions, *condexp1* and *condexp2* is:

>   condexp1 **and** condexp2
>   condexp1 **or** condexp2
>   **not** condexp1

The following example includes the **not** operator:

>   **not** $(x \leq q)$

The following example in pseudo-code includes the **or** logical operator:

>   **if** a < b **or** $x \geq y$
>   **then**
>   ⟨ block_1 ⟩
>   **else**
>   ⟨ block_2 ⟩
>   **endif**

The previous condition can also be written in the following manner:

a **less than** b **or** x **is greater or equal to** y

MATLAB and Octave provides several logical operators and functions for constructing complex conditions from simpler ones. The basic logical operators for scalars (individual numbers) in MATLAB or Octave are: the && (**and**) operator, the || (**or**) operator, and the ~ (**not**) operator.

In the following example, two simple conditions are combined with the && operator (**and**). Note that the ; is a special operator used in a statement for suppressing output of the statement. The result returned by MATLAB and Octave, ans = 1, indicates that the combined condition evaluated to *true*.

```
>> x = 5.25;
>> (2.5 < x) && (x < 21.75)
ans = 1
```

7.7 Summary

The selection structure is also known as alternation. It evaluates a condition then follows one of two (or more) paths. The two general selection statements **if** and **case** statements are explained in pseudo-code and in MATLAB and Octave. The first one is applied when there are two or more possible paths in the algorithm, depending on how the condition evaluates. The case statement is applied when the value of a single variable or expression is evaluated, and there are multiple possible values.

The condition in the **if** statement consists of a conditional expression, which evaluates to a truth-value (true or false). Relational operators and logical operators are used to form more complex conditional expressions.

Key Terms			
selection	alternation	condition	if statement
case statement	relational operators	logical operators	truth value
then	else	endif	endcase
otherwise	elseif	end	

Exercises

Exercise 7.1 Develop a computational model that computes the conversion from gallons to liters and from liters to gallons. Include a flowchart, pseudo-code design and a complete implementation in MATLAB or Octave. The user inputs the string: 'gallons' or 'liters', the model then computes the corresponding conversion.

Exercise 7.2 Develop a computational model to calculate the total amount to pay for movie rental. Include a flowchart, pseudo-code design and a complete implementation in MATLAB or Octave. The movie rental store charges $3.50 per day for every DVD movie. For every additional period of 24 hours, the customer must pay $0.75.

Exercise 7.3 Develop a computational model that finds and display the largest of several numbers, which are read from the input device. Include a flowchart, pseudo-code design and a complete implementation in MATLAB or Octave.

Exercise 7.4 Develop a computational model that finds and display the smallest of several numbers, which are read from the input device. Include a flowchart, pseudo-code design and a complete implementation in MATLAB or Octave.

Exercise 7.5 Develop a computational model program that computes the gross and net pay of the employees. The input quantities are employee name, hourly rate, number of hours, percentage of tax (use 14.5%). the tax bracket is $ 115.00. When the number of hours is greater than 40, the (overtime) hourly rate is 40% higher. Include a flowchart, pseudo-code design and a complete implementation in MATLAB or Octave.

Exercise 7.6 Develop a computational model that computes the distance between two points. A point is defined by a pair of values (x, y). Include a flowchart, pseudo-code design and a complete implementation in MATLAB or Octave. The distance, d, between two points, $P_1(x_1,y_1)$ and $P_2(x_2,y_2)$ is defined by:

$$d = \sqrt{(x_2 - x_1)^2 + (y_2 - y_1)^2}$$

Exercise 7.7 Develop a computational model that computes the fare on a ferry that transports passengers with motor vehicles. Include a flowchart, pseudo-code design and a complete implementation in MATLAB or Octave. Passengers pay an extra fare based on the vehicle's weight. Use the following data: vehicles with weight up to 780 lb pay $80.00, up to 1100 lb pay $127.50, and up to 2200 lb pay $210.50.

Exercise 7.8 Develop a computational model that computes the average student grades. The input data are the four letter grades for various work submitted by the students. Include a flowchart, pseudo-code design and a complete implementation in MATLAB or Octave.

Chapter 8

Repetition

8.1 Introduction

This chapter covers the repetition structure for specifying, describing, and implementing algorithms in computational models. The repetition structure and the corresponding statements are discussed with pseudo-code, MATLAB, and Octave. The repetition structure specifies that a block of statements be executed repeatedly based on a given condition. Basically, the statements in the block of code are executed several times, so this structure is often called a *loop* structure. An algorithm that includes repetition has three major parts in its form:

1. the initial conditions

2. the steps that are to be repeated

3. the final results.

There are three general forms of the repetition structure: the **while** loop, the **repeat-until** loop, and the **for** loop. The first form of the repetition structure, the **while** construct, is the most flexible. The other two forms of the repetition structure can be expressed with the **while** construct.

8.2 Repetition with While Construct

The **while** construct consists of a conditional expression and block of statements. This construct evaluates the condition before the block of statements is executed. If the condition is true, the statements in the block are executed. This repeats while the condition evaluates to true; when the condition evaluates to false, the loop terminates.

8.2.1 While-Loop Flowchart

A flowchart with the **while** loop structure is shown in Figure 8.1. The block of statements consists of a sequence of actions.

The actions in the block of statements are performed while the condition is true. After the actions in the block are performed, the condition is again evaluated, and the actions are again performed if the condition is still true; otherwise, the loop terminates.

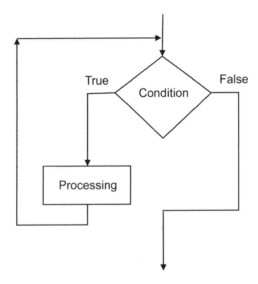

FIGURE 8.1: A flowchart with a while-loop

The condition is tested first, and then the block of statements is performed. If this condition is initially false, the statements in the block are not performed.

The number of times that the loop is performed is normally a finite number. A well-defined loop will eventually terminate, unless it has been specified as a non-terminating loop. The condition is also known as the *loop condition*, and it determines when the loop terminates. A non-terminating loop is defined in special cases and will repeat the actions forever.

8.2.2 The While Structure in Pseudo-Code

The form of the **while** statement includes the condition, the statements in the block, and the keywords **while**, **do**, and **endwhile**. The block of statements is placed after the **do** keyword and before the **endwhile** keyword. The following lines of pseudo-code show the general form of the while-loop statement that is shown in the flowchart of Figure 8.1.

```
while ⟨ condition ⟩ do
    ⟨ block of statements ⟩
endwhile
```

The following example has a **while** statement with a condition that checks the value of variable j. The block of statements repeats while the condition j <= MAX_NUM is true.

```
while j <= MAX_NUM do
    set sum = sum + 12.5
    set y = x * 2.5
    add 3 to j
endwhile
display "Value of sum: ", sum
display "Value of y: ", y
```

8.2.3 While-Loop with MATLAB and Octave

The following lines of code show the general form of the while-loop statement in MATLAB and Octave; it is similar to the pseudo-code statement and follows the loop definition shown in the flowchart of Figure 8.1.

```
while ⟨ condition ⟩
    ⟨ block of statements ⟩
end
```

The previous example has a **while** statement with a condition that checks the value of variable j. The block of statements repeats while the condition j <= MAX_NUM is true. The following lines of code show the MATLAB (and Octave) implementation.

```
while j <= MAX_NUM
    sum = sum + 12.5
    y = x * 2.5
    j = j + 3
end
moutput('Value of sum: ', sum)
moutput('Value of y: ', y)
```

8.2.4 Loop Counter

As mentioned previously, in the while-loop construct the condition is tested first and then the statements in the statement block are performed. If this condition is initially false, the statements are not performed.

The number of times that the loop is performed is normally a finite integer value. For this, the condition will eventually be evaluated to false, that is, the loop will eventually terminate. This condition is often known as the *loop condition*, and it determines when the loop terminates. Only in some very special cases, the programmer can decide to write an infinite loop; this will repeat the statements in the repeat loop forever.

A counter variable stores the number of times (also known as iterations) that the loop executes. The counter variable is incremented every time the statements in the loop are performed. The variable must be initialized to a given value, typically to 0 or 1.

In the following pseudo-code listing, there is a while statement with a counter variable with name *loop_counter*. This counter variable is used to control the number of times the block statement is performed. The counter variable is initially set to 1, and is incremented every time through the loop.

```
Max_Num = 25      // maximum number of times
                  // to execute the loop
set loop_counter = 1   // initial value of counter
while loop_counter < Max_Num do
   increment loop_counter
   display "Value of counter: ", loop_counter
endwhile
```

The first time the statements in the block are performed, the loop counter variable *loop_counter* has a value equal to 1. The second time through the loop, variable *loop_counter* has a value equal to 2. The third time through the loop, it has a value of 3, and so on. Eventually, the counter variable will have a value equal to the value of *Max_Num* and the loop terminates.

8.2.5 Accumulator Variables

An accumulator variable stores partial results of repeated calculations. The initial value of an accumulator variable is normally set to zero.

For example, an algorithm that calculates the summation of numbers from input, includes an accumulator variable. The following pseudo-code statement accumulates the values of *innumber* in variable *total*, it should be included in a while loop:

```
total = 0.0
while j < Max_num
   add innumber to total
endwhile
display "Total accumulated: ", total
```

After the **endwhile** statement, the value of the accumulator variable *total* is displayed using the pseudo-code statement to print the string "Total accumulated" and the value of *total*.

Counter and accumulator variables serve a specific purpose. These variables should be well documented.

8.2.6 Summation of Input Numbers

The following simple problem applies the concepts and implementation of while loop and accumulator variable. The problem computes the summation of numeric values input from the main input device. The computation should proceed while the input values are greater than zero.

The pseudo-code that describes the algorithm uses an input variable, an accumulator variable, a loop counter variable, and a conditional expression that evaluates whether the input value is greater than zero.

```
set innumber = 1.0   // number with dummy initial value
set loop_counter = 0
set sum = 0.0        // initialize accumulator variable
while innumber > 0.0  do
   display "Type number: "
   read innumer
   if innumber > 0.0
   then
      add innumber to sum
      increment loop_counter
      display "Value of counter: ", loop_counter
   endif
endwhile
display "Value of sum: ", sum
```

Listing 8.1 shows the commands that implement the summation problem in MATLAB and Octave. These commands are stored in the script file summ.m.

122   *Introduction to Elementary Computational Modeling*

Listing 8.1 MATLAB/Octave command file for summation.
```
1  % MATLAB/Octave script that computes summation
2  % File: summ.m
3  % number with dummy initial value
4  innumber = 1.0;
5  loop_counter = 0;
6  sum = 0.0; % initialize accumulator var
7  while innumber > 0.0
8      innumber = input('Type number: ');
9      if innumber > 0.0
10         sum = sum + innumber;
11         loop_counter = loop_counter + 1;
12         moutput('Value of counter: ', loop_counter)
13     end
14 end
15 moutput('Value of sum: ', sum)
```

The following output listing includes the execution of the MATLAB command file `summ.m`.

```
octave-3.2.4.exe:36> summ
Type number: 12.5
Value of counter: 1
Type number: 10.0
Value of counter: 2
Type number: 5.5
Value of counter: 3
Type number: 4.25
Value of counter: 4
Type number: 133.75
Value of counter: 5
Type number: 0.0
Value of sum: 166
```

8.3 Repeat-Until Construct

The **repeat-until** construct is a control flow statement that allows code to be executed repeatedly based on a given condition. This construct consists of a block of code and the condition.

The actions within the block are executed first, and then the condition is evaluated. If the condition is not true the actions within the block are executed again. This repeats until the condition becomes true.

Repeat-until structures check the condition after the block is executed, this is an important difference with while loop, which tests the condition before the actions within the block are executed. Figure 8.2 shows the flowchart for the repeat-until structure.

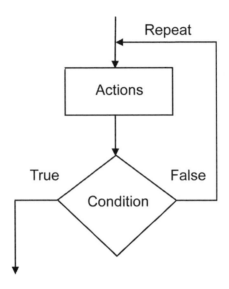

FIGURE 8.2: A flowchart with a repeat-until structure

The pseudo-code statement of the repeat-until structure corresponds directly with the flowchart in Figure 8.2 and uses the keywords **repeat**, **until**, and **endrepeat**. The following portion of code shows the general form of the repeat-until statement.

> **repeat**
>   ⟨ statements in block ⟩
> **until** ⟨ condition ⟩
> **endrepeat**

The following lines of code show the pseudo-code listing of a repeat-until statement for the previously discussed problem. MATLAB or Octave do not have an equivalent statement.

```
set innumber = 1.0   // number with dummy initial value
set loop_counter = 0
set sum = 0.0        // initialize accumulator variable
```

```
repeat
   display "Type number: "
   read innumer
   if innumber > 0.0
   then
      add innumber to sum
      increment loop_counter
      display "Value of counter: ", loop_counter
   endif
until innumber <= 0.0
endrepeat
display "Value of sum: ", sum
```

8.4 For Loop Structure

The **for** loop structure explicitly uses a loop counter; the initial value and the final value of the loop counter are specified. The **for** loop is most useful when the number of times that the loop is carried out is known in advance. In pseudo-code, the **for** statement includes the keywords: **for**, **to**, **downto**, **do**, and **endfor**. The **for** statement has the following general form in pseudo-code:

> **for** ⟨ *counter* ⟩ = ⟨ *initial_val* ⟩ **to** ⟨ *final_val* ⟩
> **do**
>   Block of statements
> **endfor**

On every iteration, the loop counter is automatically incremented. The last time through the loop, the loop counter has its final value allowed. In other words, when the loop counter reaches its final value, the loop terminates. The **for** loop is similar to the **while** loop in that the condition is evaluated before carrying out the operations in the repeat loop.

The following portion of pseudo-code uses a **for** statement for the repetition part of the summation problem.

```
for j = 1 to num do
   set sum = sum + 12.5
   set y = x * 2.5
endfor
display "Value of sum: ", sum
display "Value of y: ", y
```

Variable *j* is the counter variable, which is automatically incremented and is

Repetition

used to control the number of times the statements in a block will be performed. In MATLAB and Octave, the previous portion of code is written as follows.

```
for j = 1:num
   sum = sum + 12.5
   y = x * 2.5
end
moutput('Value of sum: ', sum)
moutput('Value of y: ', y)
```

8.4.1 The Summation Problem with a For Loop

Using the for-loop construct of the repetition structure, the algorithm for the summation of input data can be defined in a relatively straightforward manner with pseudo-code. The most significant differences from the previous design is that the number of input data to read from the input device is included at the beginning of the algorithm.

```
set innumber = 1.0   // number with dummy initial value
set sum = 0.0        // initialize accumulator variable
display "Number of input data to read: "
read MaxNum
for loop_counter = 1 to MaxNum do
   display "Type number: "
   read innumer
   if innumber > 0.0
   then
      add innumber to sum
      display "Value of counter: ", loop_counter
   endif
endfor
display "Value of sum: ", sum
```

Listing 8.2 shows the MATLAB/Octave commands that implement the summation problem with a for-loop. These commands are stored in the `summfor.m` script file.

Listing 8.2 MATLAB/Octave script for summation with for-loop.

```
1 % MATLAB/Octave command file to compute summation
2 % using a for-loop
3 % File: summfor.m
4 % number with dummy initial value
5 innumber = 1.0
6 sum = 0.0   % initialize accumulator var
7 MaxNum = input('Type number of input data: ');
```

```
 8  for loop_counter 1:MaxNum
 9     innumber = input('Type number: ');
10     if innumber > 0.0
11         sum = sum + innumber;
12         moutput('Value of counter: ', loop_counter)
13     end
14  end
15  moutput('Value of sum: ', sum)
```

8.4.2 The Factorial Problem

The *factorial* operation, denoted by the symbol !, can be defined in a general and informal manner as follows:

$$y! = y \times (y-1) \times (y-2) \times (y-3) \times \ldots \times 1$$

For example, the factorial of 5 is:

$$5! = 5 \times 4 \times 3 \times 2 \times 1$$

8.4.2.1 Mathematical Specification of Factorial

A mathematical specification of the factorial function is as follows, for $y \geq 0$:

$$y! = \begin{cases} 1, & \text{when } y = 0 \\ y \times (y-1)!, & \text{when } y > 0 \end{cases}$$

The base case in this definition is the value of 1 for the function if the argument has value zero, that is $0! = 1$. The general (recursive) case is $y! = y \times (y-1)!$, if the value of the argument is greater than zero. This function is not defined for negative values of the argument.

8.4.2.2 Computing Factorial

The factorial function, named *fact*, has one argument: the value for which the factorial is to be calculated. For example, $y!$ is denoted as *fact(y)*.

The MATLAB (and Octave) commands that implement function *fact(y)* are stored in the command file `fact.m` and is shown in Listing 8.3.

Listing 8.3 MATLAB/Octave command file for computing factorial.
```
1  function output = fact(y)
2  % MATLAB/Octave script file that computes factorial
3  % as a function
4  % File: fact.m
5  f = 1;
6  % Use a for-loop to compute the factorial of y
7  % Factorial is the product of integer numbers from 1 to y
8  % First check if the argument is greater than zero
9  if y > 0
10     for n = 1:y
11        f = f*n;
12     end
13     output = f;
14 elseif y == 0
15     output = 1;
16 else                    % if argument is negative
17     output = -1;
18 end
```

Note that this implementation will always return -1 for negative values of the argument. The following lines of MATLAB commands invoke or call the *fact(y)* function with various values of the argument.

```
octave-3.2.4.exe:38> fact(5)
ans = 120
octave-3.2.4.exe:39> fact(8)
ans =    4032
octave-3.2.4.exe:40> fact(10)
ans =    36288
octave-3.2.4.exe:41> fact(-5)
ans = -1
```

8.5 Summary

The repetition structures is used in algorithms in order to perform repeatedly a group of steps (instructions) in the block of statements. There are three types of loop structures: **while** loop, **repeat-until**, and **for** loop. In the **while** construct, the loop condition is tested first, and then the block of statements is performed if the condition is true. The loop terminates when the condition is false.

In the **repeat-until** construct, the group of statements in the block is carried out first, and then the loop condition is tested. If the loop condition is true, the loop terminates; otherwise the statements in the block are performed again.

The number of times the statements in the block is carried out depends on the condition of the loop. In the **for** loop, the number of times to repeat execution is explicitly indicated by using the initial and final values of the loop counter. Simple examples are executed in MATLAB and Octave.

Key Terms			
repetition	loop	while	loop condition
do	endrepeat	block	loop termination
loop counter	endwhile	accumulator	repeat-until
for	to	downto	endfor
end	iterations	summation	factorial

Exercises

Exercise 8.1 Develop a computational model that computes the maximum value from a set of input numbers. Use a while loop in the algorithm and implement in MATLAB or Octave.

Exercise 8.2 Develop a computational model that computes the maximum value from a set of input numbers. Use a for loop in the algorithm and implement in MATLAB or Octave.

Exercise 8.3 Develop a computational model that computes the maximum value from a set of input numbers. Use a repeat-until loop in the algorithm and implement in MATLAB or Octave.

Exercise 8.4 Develop a computational model that finds the minimum value from a set of input values. Use a while loop in the algorithm and implement in MATLAB or Octave.

Exercise 8.5 Develop a computational model that finds the minimum value from a set of input values. Use a for loop in the algorithm and implement in MATLAB or Octave.

Exercise 8.6 Develop a computational model that finds the minimum value from a set of input values. Use a repeat-until loop in the algorithm and implement in MATLAB or Octave.

Exercise 8.7 Develop a computational model that computes the average of a set of input values. Use a while loop in the algorithm and implement in MATLAB or Octave.

Exercise 8.8 Develop a computational model that computes the average of a set of input values. Use a for loop in the algorithm and implement in MATLAB or Octave.

Exercise 8.9 Develop a computational model that computes the average of a set of input values. Use a repeat-until loop in the algorithm and implement in MATLAB or Octave.

Exercise 8.10 Develop a computational model that computes the student group average, maximum, and minimum grade. The computational model uses the input grade for every student. Use a while loop in the algorithm and implement in MATLAB or Octave.

Exercise 8.11 Develop a computational model that computes the student group average, maximum, and minimum grade. The computational model uses the input grade for every student. Use a for loop in the algorithm and implement in MATLAB or Octave.

Exercise 8.12 Develop a computational model that reads rainfall data in inches for yearly quarters from the last five years. The computational model is to compute the average rainfall per quarter, the average rainfall per year, and the maximum rainfall per quarter and for each year. Implement in MATLAB or Octave.

Exercise 8.13 Develop a computational model that computes the total inventory value amount and total per item. The computational model is to read item code, cost, and description for each item. The number of items to process is not known. Implement in MATLAB or Octave.

Chapter 9

Data Lists

9.1 Introduction

Most algorithms use data lists or arrays, each one of which can store a large number of values in a single collection. A data list or *array* can store multiple values each of which can be referred to with the same name. The individual values in the array are known as *elements*.

Programming languages such as Ada, Eiffel, C++, and Java support arrays to handle large numbers of values in data collections. MATLAB and Octave can manipulate individual elements of arrays and provide functions that manipulate entire arrays with a single operation. To access a particular element of the array, an integer value or variable known as *index* is used with the name of the array. The values of the index represent the relative position of the element in the array. Figure 9.1 shows a simple array with 13 elements.

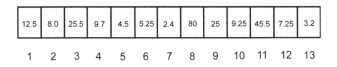

FIGURE 9.1: A simple array.

Algorithms that define and manipulate arrays normally include the following sequence of steps:

1. Define or declare the array with appropriate name, size, and type
2. Create the array
3. Assign initial values to the array elements
4. Access or manipulate the individual elements of the array

In MATLAB and Octave, an array is created as a *matrix*, which is typically a multidimensional array. The most common form of an array is a double-dimensional array or matrix, which is an array with data items organized into rows and columns. In this case, two indexes are used, one index to indicate the column and one index to indicate the row. Figure 9.2 shows a matrix with 13 columns and two rows.

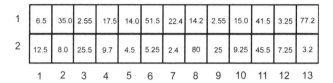

FIGURE 9.2: A simple two-dimensional array.

9.2 Creating an Array

An array is created with an appropriate size and the values of the various elements. The size of the array is the number of elements it can store.

9.2.1 Creating Arrays in Pseudo-Code

In pseudo-code, the general form for creating an array is:

> **create** ⟨ array_name ⟩ **array** [⟨ size ⟩]
> **of type** ⟨ array_type ⟩

The following pseudo-code statement creates array y, with a capacity of 20 elements.

```
create y array [20] of type float
```

The size of the array can be specified using a symbolic (identifier) constant. For example, given the constant *MAX_NUM* with value 20, the array y can be created in pseudo-code:

```
create y array [MAX_NUM] of type float
```

9.2.2 Creating Arrays in MATLAB and Octave

In MATLAB and Octave, an array is created dynamically when initially assigning values to its various elements. A vector is a single-dimension array and is very common in MATLAB and Octave. There are three basic techniques to create arrays in MATLAB and Octave.

The simplest way to create a vector in MATLAB and Octave is to use an assignment operator, the name of the vector, and the various values enclosed in brackets. For example, assume that the name of the vector shown in Figure 9.1 is *precip*; it can be created in MATLAB and Octave in the following manner:

```
octave-3.2.4.exe:45>
  precip = [12.5 8.0 25.5 9.7 4.5 5.25 2.4 80 25 9.25 45.5
            7.25 3.2]
```

Recall that a vector is a single-dimension array and by default is taken as a row vector, which a special case of a matrix. A row vector is a matrix of one row and multiple columns. After entering the previous line at the MATLAB prompt, MATLAB responds with the following lines:

```
    precip =
      Columns 1 through 4
      12.5000   8.0000   25.5000   9.70
      Columns 5 through 8
       4.5000   5.2500    2.4000  80.00
      Columns 9 through 12
      25.000    9.2500   45.500    7.2500
      Column 13
       3.2000
```

The second technique to create a single-dimension array (vector) in MATLAB and Octave is using equally spaced values and fixed increments. The special notation *(init_val:increment:last_val)* is used in an assignment to the vector variable.

For example, the following line of code in MATLAB creates a vector with name *y*, a starting value of 2.0, increments of 0.5, and a final value of 13.5.

```
    octave-3.2.4.exe:47> y = (2.0:0.5:13.5)
```

MATLAB immediately responds by printing the values in the array.

```
y =
  Columns 1 through 4
    2.0000    2.5000    3.0000    3.50
  Columns 5 through 8
    4.0000    4.5000    5.0000    5.50
  Columns 9 through 12
    6.0000    6.5000    7.0000    7.50
  Columns 13 through 16
    8.0000    8.5000    9.0000    9.50
  Columns 17 through 20
   10.0000   10.5000   11.0000   11.50
  Columns 21 through 24
   12.0000   12.5000   13.0000   13.50
```

The third technique to create a single-dimension array in MATLAB and Octave uses the function *linspace* in an assignment to the vector variable. This also creates a vector with the values equally spaced by invoking the function *linspace* using the initial value, the final value, and the number of values in the vector.

For example, to create the same vector as in the previous example, the MATLAB and Octave function *linspace* is used in the following manner.

```
octave-3.2.4.exe:48> y = linspace(2.0, 13.5, 24)
```

MATLAB and Octave immediately respond by printing the 24 element values in the array. MATLAB and Octave also have slightly more complex facilities to create arrays from existing ones. For example, the following MATLAB and Octave statement creates a vector z that includes the values of vector y and two additional element values 10.24 and 61.5.

```
octave-3.2.4.exe:49> z = [y 10.24 61.5]
```

Vector z is created with the new element values 10.24 and 61.5 as the values of the last two elements. The following listing shows the values in vector z.

```
z =
  Columns 1 through 4
    2.0000    2.5000    3.0000    3.5000
  Columns 5 through 8
    4.0000    4.5000    5.0000    5.5000
  Columns 9 through 12
    6.0000    6.5000    7.0000    7.5000
  Columns 13 through 16
    8.0000    8.5000    9.0000    9.5000
  Columns 17 through 20
   10.0000   10.5000   11.0000   11.5000
  Columns 21 through 24
   12.0000   12.5000   13.0000   13.5000
  Columns 25 through 26
   10.2400   61.5000
```

The following lines of code, create a row vector named *w* with element values starting with 1 up to 9 in increments of 2. This row vector is used to construct a second vector named *y*, with the elements of *w* followed by five new element values taken from a starting value of 2 ending in 11 in increments of 2.

```
octave-3.2.4.exe:56> w = [1:02:9]
w =
   1   3   5   7   9
octave-3.2.4.exe:57> y = [w 2:2:11]
y =
   1   3   5   7   9   2   4   6   8   10
```

9.3 Operations on Arrays

MATLAB and Octave support a wide variety of operations on arrays; the two group of operations that can be performed on arrays are.

1. Operations with individual elements of the array

2. Operations on the entire array

An algorithm can manipulate an array by accessing the individual elements of the array and performing some computation. Recall that an integer value known as the *index* is used with the name of the array to access an individual element of an array. In MATLAB and Octave, the range of index values starts with 1 and ends with the number equal to the size of the array.

9.3.1 Array Elements in Pseudo-Code

In pseudo-code, a particular element of an array is accessed using the name of the array followed by the index value in rectangular brackets. The following statement assigns a value of 14.5 to element 5 of array *y*. In most notations or programming languages, the index values start with 0.

```
set y[4] = 14.5
```

The index used can be an integer constant, a symbolic constant (identifier), or an integer variable. The following statement assigns the value 94.5 to element 6 of array *height*, using variable *k* as the index.

```
create height array [MAX_NUM] of type float
set k = 5
set height[k] = 94.5
```

9.3.2 Using Array Elements with MATLAB and Octave

Accessing the elements of a vector in MATLAB and Octave is also known as indexing a vector. Accessing a particular element of an array uses the name of the array followed by the index value in parenthesis. The following assignment statement assigns a value of 14.5 to element 5 of array *y*.

```
y(5) = 14.5
```

The index used can be an integer constant, a symbolic constant (identifier), or an integer variable. The following lines of code in MATLAB and Octave create vector *height*, assign a value 6 to variable *k*, and assign the value 94.5 to element 6 of array *height*, using variable *k* as the index.

```
octave-3.2.4.exe:50> height = [12.5 8.25 7.0 21.35 6.55]
octave-3.2.4.exe:51> k = 6
octave-3.2.4.exe:52> height(k) = 94.5
```

MATLAB and Octave respond by changing the value of element 6 in vector *height*.

```
height =
  Columns 1 through 4
  12.5000    8.2500    7.0000    21.3500
  Columns 5 through 6
   6.5500   94.5000
```

Data Lists

In addition to accessing individual elements of an array in a simple assignment statement, more complex arithmetic operations are also supported in MATLAB and Octave. The following example involves a trigonometric operation with element 5 of vector *height*; the assignment gives a value to variable *a*. The MATLAB and Octave symbolic constant *pi* is the value for π.

```
octave-3.2.4.exe:53> a = sin(0.04*height(5)/pi)
a =
   0.08330055128050
```

To get the size of a vector, MATLAB and Octave provide the *length* function, which returns an integer value that represents the number of elements in the vector. The following line of code assigns to variable *n* the size of vector *height*.

```
octave-3.2.4.exe:55>  n = length(height)
n =
   6
```

9.3.3 Arithmetic Operations on Vectors

Simple arithmetic operations can be performed between a vector and a number, known as a *scalar*. These operations are: addition, subtraction, multiplication, and division. These operations are applied to all elements of the vector. For example, the following command adds the scalar 3.25 to all elements of vector *height*, which was previously created. The assignment statement creates a new vector *q*.

```
octave-3.2.4.exe:54> q = height + 3.25
q =
  Columns 1 through 4
   15.7500   11.5000   10.2500   24.6000
  Columns 5 through 6
    9.8000   97.7500
```

Several array operations are supported in MATLAB and Octave. The simplest array operations are the element by element operations on entire vectors, these are: addition, subtraction, multiplication, division, and exponentiation of two vectors. Element by element operations use the dot (.) notation for multiplication, division, and exponentiation.

The following lines of code in MATLAB and Octave apply several simple array operations on two small row vectors.

```
>> x = [1 2 3]
x =
     1     2     3
>> y = [4 5 6]
y =
     4     5     6
>> x + y
ans =
     5     7     9
>> y - x
ans =
     3     3     3
>> z = x .* y
z =
     4    10    18
```

Without using the dot notation in multiplication of two row vectors of the same size, MATLAB and Octave attempt to perform matrix multiplication instead of element by element multiplication, and MATLAB and Octave respond with an error message.

9.4 Multidimensional Arrays

A multidimensional array is created with more than one dimension. As mentioned previously, two-dimensional arrays are also known as matrices and are mathematical structures with values arranged in columns and rows. Two index values are required, one for the rows and one for the columns.

9.4.1 Multidimensional Arrays with Pseudo-Code

To create a double-dimensional array in pseudo-code, two numbers are used. The first number defines the range of values for the first index (rows) and the second number defines the range of values for the second index (columns).

The following pseudo-code statements create a two-dimensional array named *precip* with size 30 rows and 10 columns.

```
define ROWS = 30 of type integer
define COLS = 10 of type integer
...
create precip array [ROWS][COLS] of type float
```

To reference individual elements of a two-dimensional array, two integer numbers are used as indexes. In the following pseudo-code statements each individual element is accessed using index variables *i* and *j*. The code assigns all the elements of array *precip* to 0.0:

```
for j = 0 to COLS - 1 do
   for i = 0 to ROWS - 1 do
      set precip [i][j] = 0.0
   endfor
endfor
```

Nested loops are used, an outer loop and an inner loop. The inner loop varies the row index *i* from 0 to value *ROWS-1*, and the outer loop varies the row index *j* from 0 to *COLS-1*. The assignment sets the value 0.0 to the element at row *i* and column *j*.

9.4.2 Multidimensional Arrays with MATLAB and Octave

To create a two-dimensional array in MATLAB and Octave, the semicolon (;) is used to separate the rows in the array (or matrix). The following lines of code create a row vector named *x* with element values taken from a starting value of 2 ending in 11 in increments of 2. Using the semicolon (;) symbol, a two-dimensional array named *z* is created with the first row with the elements of row vector *w*. Using the semicolon, a second row is defined from the values in vector *x*.

```
octave-3.2.4.exe:56> w = [1:02:9]
w =
     1     3     5     7     9
octave-3.2.4.exe:58> x = [2:2:11]
x =
     2     4     6     8    10
octave-3.2.4.exe:59> z = [w; x]
z =
     1     3     5     7     9
     2     4     6     8    10
```

To access individual elements of a two-dimensional array, two indexes are used. The first number is the index for rows and the second number is the index for columns. To access the value of the element in array *z* defined previously of the element at row 2 and column 3, the notation is: $z(2,3)$. For example:

```
> zz = z(2,3) + 23.5
```

A technique to create a matrix that has some elements with unknown values uses the function *zeros*. This creates a matrix of a given size with the elements having a zero value. Specific elements of the matrix can then be assigned values. The following example creates a matrix with two rows and five columns, then the value of the elements of the first row are changed to 1.25.

```
octave-3.2.4.exe:60> a = zeros(2,5)
a =
   0   0   0   0   0
   0   0   0   0   0
octave-3.2.4.exe:61> for j=1:5
   a(1,j) = 1.25;
end
a =
   1.2500   1.2500   1.2500   1.2500   1.2500
        0        0        0        0        0
```

The following lines of code create a two-dimensional array named *precip* with size 2 rows and 5 columns and assign the value 1.5 to every element of the array.

```
    COLS = 5
    ROWS = 2
    precip = zeros(ROWS, COLS)
    for j = 1:COLS
        for i = 1:ROWS
            precip(i,j) = 1.5
        end
    end
```

This code was stored as a script with name *double_dimen* (the file name is double_dimen.m). Executing the script at the MATLAB (and Octave) prompt gives the following output.

```
octave-3.2.4.exe:62> double_dimen
COLS =
    5
ROWS =
    2
precip =
    0   0   0   0   0
    0   0   0   0   0
precip =
    1.5000   1.5000   1.5000   1.5000   1.5000
    1.5000   1.5000   1.5000   1.5000   1.5000
```

9.5 Applications Using Arrays

This section discusses several simple applications of arrays; a few of these applications perform simple manipulation of arrays, other applications perform slightly more complex operations with arrays such as searching and sorting.

9.5.1 Problems with Simple Array Manipulation

The problems discussed in this section compute the average value and the maximum value in an array named *varr*. The algorithms that solve these problems examine all the elements of the array.

9.5.1.1 The Average Value in an Array

To compute the average value in an array, the algorithm is designed to first compute the summation of all the elements in the array, the accumulator variable *sum* is used to store this. Second, the algorithm computes the average value by diving the value of *sum* by the number of elements in the array. The following listing has the pseudo-code description of the algorithm.

1. Initialize the value of the accumulator variable, *sum*, to zero.
2. For every element of the array, add its value to the accumulator variable *sum*.
3. Divide the value of the accumulator variable by the number of elements in the array, *num*.

The accumulator variable *sum* stores the summation of the element values in the array named *varr* with *num* elements. The average value, *ave*, of array *varr* using index j starting with $j = 1$ to $j = n$ is expressed mathematically as:

$$ave = \frac{1}{num} \sum_{j=1}^{num} varr_j$$

The following listing describes the algorithm with pseudo-code. The size of the array (*num*) and the array elements are entered from the input device.

```
description
   This algorithm computes the average of the
   elements in array varr.
   */
variables
   define sum of type float
```

Introduction to Elementary Computational Modeling

```
   define ave of type float    // average value
   define j of type integer
begin
   display "Enter array size"
   read num
   for j = 0 to num - 1 do
     display "Enter element value: "
   endfor
   //
   set sum = 0
   for j = 0 to num - 1 do
     add myarr[j] to sum
   endfor
   set ave = sum / num
   display "Average value in array: ", ave
```

Listing 9.1 shows the MATLAB and Octave commands that implement the algorithm that computes the average value of the elements in the array. This code is stored in the script file `aver.m`.

Listing 9.1: MATLAB/Octave command file for computing average.

```
1  % MATLAB/Octave command file to compute average
2  % This script inputs the array size
3  % and the elements of the array from the console
4  % Computes the average value in the array
5  % File: aver.m
6  num = input('Enter array size: ');
7  for j=1:num
8      varr(j) = input('Enter array element: ');
9  end
10 %
11 % Now the average value in array
12 sum = 0.0;
13 for j = 1:num
14     sum = sum + varr(j);
15 end
16 ave = sum/num;
17 moutput('Average value: ', ave)
```

The following output listing shows the result of executing the command file (script) *aver* at the MATLAB prompt.

Data Lists

```
octave-3.2.4.exe:65> aver
Enter array size: 4
Enter array element: 12.4
Enter array element: 2.45
Enter array element: 23.21
Enter array element: 22.1
Average value: 15.04
```

9.5.1.2 Maximum Value in an Array

Consider a problem that deals with finding the maximum value in an array named *varr*. The algorithm with the solution to this problem also examines all the elements of the array.

The variable *max_arr* stores the maximum value found so far. The name of the index variable is *j*. The algorithm description is:

1. Read the value of the array size, *num*, and the value of the array elements.

2. Initialize the variable *max_arr* that stores the current largest value found (so far). This initial value is the value of the first element of the array.

3. Initialize the index variable (value zero).

4. For each of the other elements of the array, compare the value of the next array element; if the value of the current element is greater than the value of *max_arr* (the largest value so far), change the value of *max_arr* to this element value, and store the index value of the element in variable *k*.

5. The index value of variable *k* is the index of the element with the largest value in the array.

The algorithm description in pseudo-code appears in the following listing.

```
description
   Find the element with the maximum
   value in the array and write its index value.   */
   variables
      define j of type integer
      define k of type integer       // index of max elelemnt
      define max_arr of type float   // largest element
   begin
      display "Enter array size: "
      read num
      for j = 0 to num-1 do
         display "Enter array element: "
         read varr[j]
```

144   *Introduction to Elementary Computational Modeling*

```
      endfor
      set k = 0
      set max_arr = varr[0]
      for j = 1 to num - 1 do
         if varr[j] > max_arr
         then
              set k = j
              set max_arr = varr[j]
         endif
      endfor
   display "index of max value: ", k
```

Listing 9.2 contains the MATLAB and Octave commands that implement the algorithm for finding the maximum value in an array; these commands are stored in the script file arrmax.m.

Listing 9.2: MATLAB/Octave command file for finding maximum value.

```
1  % MATLAB/Octave command file for finding
2  % the maximum value in an array
3  % File: arrmax.m
4  % This script inputs the array size
5  % and the elements of the array from the console
6  % Computes the maximum value in the array
7  num = input('Enter array size: ');
8  for j=1:num
9     varr(j) = input('Enter array element: ');
10 end
11 %
12 % Now find the maximum value in array
13 k = 0;
14 max_arr = varr(1);
15 for j = 1:num
16    if varr(j) > max_arr
17       k = j;
18       max_arr = varr(j);
19    end
20 end
21 moutput('index of max value: ', k)
22 moutput('Max value: ', varr(k))
```

Executing the command file arrmax.m with MATLAB produces the following output.

```
octave-3.2.4.exe:67> arrmax
Enter array size: 4
Enter array element: 12.5
Enter array element: 22.2
Enter array element: 23.45
Enter array element: 23.12
index of max value: 3
Max value: 23.45
```

9.5.2 Searching

Looking for an array element with a particular value is called searching and involves examining some or all elements of an array. The search ends when and if an element of the array has a value equal to the requested value. Two general techniques for searching are: linear search and binary search.

9.5.2.1 Linear Search

A linear search algorithm examines the elements of an array in a *sequential* manner. The algorithm examines the first element of the array, then the next element and so on until the last element of the array. Every array element is compared with the requested or key value, and if an array element is equal to the requested value, the algorithm has found the element and the search terminates. This may occur before the algorithm has examined all the elements of the array.

The result of this search is the index of the element in the array that is equal to the requested value. If the requested value is not found, the algorithm indicates this with a negative result or in some other manner. The following is an algorithm description of a general linear search using a search condition of an element equal to the value of a *key*.

1. Repeat for every element of the array:

    (a) Compare the current element with the requested value or key. If the value of the array element satisfies the condition, store the value of the index of the element found and terminate the search.

    (b) If values are not equal, continue search.

2. If no element with value equal to the value requested is found, set the result to value -1.

The following algorithm in pseudo-code searches the array for an element with the requested value, *kval*. The algorithm outputs the index value of the element equal to the requested (or key) value *kval*. If no element with value equal to the requested value is found, the algorithm outputs a negative value.

```
description
  This algorithm performs a linear search of array varr using
  key value kvar. The result is the index value of the element
  found, or -1 */
variables
    define kval of type float,
    define varr array [] of type float,
    define num of type integer
    define j of type integer
    define found = false of type boolean
  begin
    set j = 0
    while j < num and found not equal true do
      if varr [j] == kval
      then
          set result = j
          set found = true
      else
          increment j
      endif
    endwhile
    if found not equal true
    then
        set result = -1
    endif
  end
```

The algorithm outputs the index value of the element that satisfies the search condition, whose value is equal to the requested value *kval*. If no element is found that satisfies the search condition, the algorithm outputs a negative value. The code is stored in the script file lsearch.m.

The following listing shows the execution of the script file marray that creates an array.

```
octave-3.2.4.exe:68> marray
Enter array size: 4
Enter array element: 12.5
Enter array element: 10.25
Enter array element: 14.585
Enter array element: 11.25
octave-3.2.4.exe:59> lsearch
Enter the key value: 11.25
Result is: 4
```

The script file lsearch.m performs the linear search in Octave. Listing

9.3 shows the MATLAB and Octave commands that implement the algorithm that searches the array for an element with the requested value, *kval*.

Listing 9.3: Command file for computing linear search in array.

```
1  % MATLAB/Octave command file for linear search
2  % it performs a linear search of the array varr
3  % looking for the value kvar
4  %     The algorithm sets the result, the index value of
5  %     the element found, or -1 if not found.
6  % File: lsearch.m
7  kval = input('Enter the key value: ');
8  found = 0;  % 0 is false and 1 is true
9  j = 1;
10 while j <= num & found ~= 1
11         if varr (j) == kval
12             result = j;
13             found = 1;    % found
14             moutput('Result is: ', result)
15         else
16             j = j + 1;
17         end
18 end
19 if found ~= 1
20     result = -1;
21     moutput('Result is: ', result)
22 end
```

9.5.2.2 Binary Search

Binary search is a more complex search method, compared to linear search. This search technique is very efficient compared to linear search because the number of comparisons is smaller.

A prerequisite for the binary search technique is that the element values in the array to be searched be sorted in ascending order. The array elements to include are split into two halves or partitions of about the same size. The middle element is compared with the key (requested) value. If the element with this value is not found, the search is continued on only one partition. This partition is again split into two smaller partitions until the element is found or until no more splits are possible because the element is not found.

With a search algorithm, the efficiency of the algorithm is determined by the number of compare operations with respect to the size of the array. The average number of comparisons with linear search for an array with N elements is $N/2$, and if the element is not found, the number of comparisons is N. With binary search, the number of comparisons is $\log_2 N$. The informal description of the algorithm is:

1. Assign the lower and upper bounds of the array to *lower* and *upper*.

2. While the lower value is less than the upper value, continue the search.

    (a) Split the array into two partitions. Compare the middle element with the key value.

    (b) If the value of the middle element is equal to the key value, terminate search and the result is the index of this element.

    (c) If the key value is less than the middle element, change the upper bound to the index of the middle element minus 1. Continue the search on the lower partition.

    (d) If the key value is greater or equal to the middle element, change the lower bound to the index of the middle element plus 1. Continue the search on the upper partition.

3. If the key value is not found in the array, the result is -1.

The following listing is the pseudo-code description of the binary search algorithm on array *varr*.

```
description
  This implements a binary search of array varr using
     key value kval. The result, tempinx, is the index
     value of the element found, or -1 if not found.  */
variables
        define kval of type float
        define varr array [] of type float
        define num of type integer
        define tempinx of type integer
        define found = false of type boolean
        define lower of type integer // index lower bound
        define upper of type integer   // index upper bound
        define middle of type integer  // index of middle
begin
        set lower = 0
        set upper = num
        while lower < upper and found not equal true
        do
          set middle = (lower + upper) / 2
          if kval == varr[middle]
          then
              set found = true
              set tempinx = middle  // result
          else
             if kval < myarr[middle]
```

Data Lists

```
            then
                set upper = middle - 1
            else
                set lower = middle + 1
            endif
         endif
      endwhile
      if found not equal true
      then
         set tempinx = -1 // result
      endif
end
```

Listing 9.4 shows the MATLAB and Octave commands that implement the binary search algorithm. These commands are stored in command file bsearch.m.

Listing 9.4: Command file for computing binary search in array.

```
 1 % MATLAB/Octave command file
 2 % It implements a binary search of array varr using
 3 %    key value kval. The result, tempinx, is the index
 4 %    value of the element found, or -1 if not found.
 5 % File: bsearch.m
 6 varr = [9.45 10.50 13.95 61.75 72.04 82.66 89.45]
 7 num = length(varr);
 8 kval = input('Enter key value: ');
 9 lower = 1;
10 upper = num; % array size
11 found = 0;
12 while lower <= upper & found ~= 1
13    middle = (lower + upper) / 2;
14    if kval == varr(middle)
15         found = 1;
16         tempinx = middle;  % result
17    else
18        if kval < varr(middle)
19            upper = middle - 1;
20        else
21            lower = middle + 1;
22        end
23    end
24 end
25 if found ~= 1
26     tempinx = -1 % result;
27 end
28 moutput('Result: ', tempinx)
```

Executing the *bsearch* script in MATLAB is shown in the following output listing.

```
octave-3.2.4.exe:69> bsearch
varr =
    9.4500   10.5000   13.9500   61.7500   72.0400   82.6600
   89.4500
Enter key value: 13.95
Result: 3
```

9.6 Average and Instantaneous Rate of Change

Recall that a mathematical function defines the relation between two (or more) variables. This relation is expressed as: $y = f(x)$. In this expression, variable y is a function of variable x. If the dependent variable y does not have a linear relationship with the variable x, then the graph that represents the relationship between y and x is a curve instead of a straight line.

9.6.1 Average Rate of Change

As explained previously, the average rate of change of a variable, y, with respect to variable, x, (the independent variable) is defined over a finite interval, Δx. The average rate of change is denoted as $\Delta y / \Delta x$.

The graphical interpretation of the average rate of change of a variable with respect to another is the **slope of a line** drawn in the Cartesian plane. The vertical axis is usually associated with the values of the dependent variable, y, and the horizontal axis is associated with the values of the independent variable, x.

Figure 9.3 shows a straight line on the Cartesian plane. Two points on the line, P_1 and P_2, are used to compute the slope of the line. Point P_1 is defined by two coordinate values (x_1, y_1) and point P_2 is defined by the coordinate values (x_2, y_2). The horizontal distance between the two points, Δx, is computed by the difference $x_2 - x_1$. The vertical distance between the two points is denoted by Δy and is computed by the difference $y_2 - y_1$.

The *slope* of the line is the inclination of the line and is computed by the expression $\Delta y / \Delta x$, which is the same as the average rate of change of a variable y over an interval Δx. Note that the slope of the line is constant, on any pair of points on the line.

The average rate of change of a variable y with respect to variable x over an interval Δx, is calculated between two points, P_1 and P_2. The line that connects

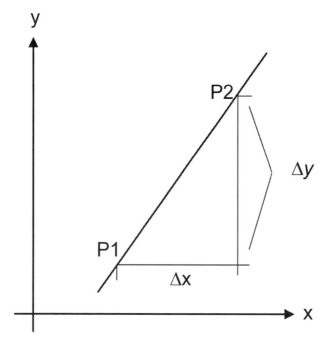

FIGURE 9.3: The slope of a line.

these two points is called a **secant** of the curve. The average rate on that interval is defined as the slope of that secant. Figure 9.4 shows a secant to the curve at points P_1 and P_2. The rate of change of y with respect to x is expressed as follows:

$$\frac{\Delta y}{\Delta x} = \frac{y_2 - y_1}{x_2 - x_1}$$

9.6.2 Instantaneous Rate of Change

The **instantaneous rate of change** of a variable, y, with respect to another variable, x, is the value of the rate of change of y at a particular value of x. This is calculated as the slope of a line that is a tangent to the curve at a point P.

Figure 9.5 shows a tangent of the curve at point P_1. The instantaneous rate of change at a specified point P1 of a curve can be approximated by calculating the slope of a secant and using a very small interval, in different words, choosing Δx very small. This can be accomplished by selecting a second point on the curve closer and closer to point P1 (in Figure 9.5), until the secant almost becomes a tangent to the curve at point P1.

Examples of average rate of change are: the average velocity, $\bar{v}$, computed

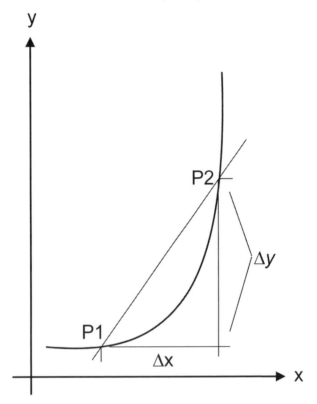

FIGURE 9.4: The slope of a secant.

by $\Delta y/\Delta t$, and the average acceleration $\bar{a}$, computed by $\Delta v/\Delta t$. These are defined over a finite time interval, Δt.

As mentioned previously, if the dependent variable y does not have a linear relationship with the variable x, then the graph that represents the relationship between y and x is a curve instead of a straight line. In this case, the instantaneous rate of change of y with respect to x is calculated for every value of y and x.

9.6.3 Computing the Rates of Change

MATLAB and Octave have the library function *diff()* that computes the differences in an array. These differences are found in a new array that has a length of 1 less than the original array. In the following lines of code, array y of length 7 is used as the argument from which to calculate the differences using function *diff()*. The new array *dy* has a length of 6.

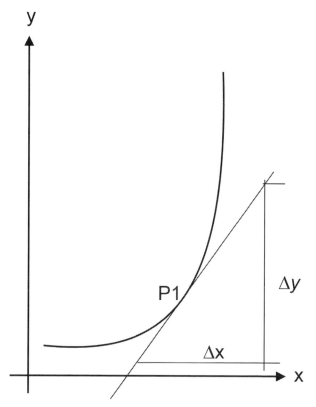

FIGURE 9.5: The slope of a tangent.

```
>> y = [9.45 10.50 13.95 61.75 72.04 82.66 89.45]
y =
    9.4500    10.5000    13.9500    61.7500    72.0400    82.6600
   89.4500
>> dy = diff(y)
dy =
    1.0500     3.4500    47.8000    10.2900    10.6200     6.7900
```

For the example of the free-falling object discussed in Chapter 5, Δt is a **finite interval** of time defined by the final time instance minus the initial time instance $(t_f - t_i)$ of the interval. A change of (vertical) displacement, Δy, is the difference $(y_f - y_i)$ in the vertical positions of the object over the time interval Δt, and a change of velocity, Δv, is the difference $(v_f - v_i)$ in the velocities in that time interval.

The velocity, $\bar{v}$, is the rate of change of the displacement with respect to time on the interval Δt, and is computed using $\Delta y/\Delta t$. The acceleration, $\bar{a}$, is

154   Introduction to Elementary Computational Modeling

the rate of change of velocity with respect to time on the interval Δt, and is computed using $\Delta v / \Delta t$.

The following listing shows the results of executing the command file rfallobj.m. The commands compute the velocity and acceleration of the free-falling object, starting with the values of the height of the object. The main concepts applied are: the velocity of the object is the rate of change of the height, and the acceleration is the rate of change of the velocity.

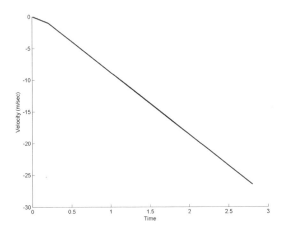

FIGURE 9.6: The velocity of the free-falling object.

```
>> rfallobj
Height of free-falling object
hf =
  Columns 1 through 7
    40.00     39.80     39.21    38.236    36.86    35.10    32.94
  Columns 8 through 14
   30.396    27.45     24.12     20.40    16.28    11.77     6.87
  Column 15
    1.5840
Differences of height
dhf =
  Columns 1 through 7
   -0.196    -0.59    -0.98     -1.37    -1.76   -2.156    -2.55
  Columns 8 through 14
    -2.94    -3.33    -3.72     -4.12   -4.508    -4.90    -5.29
Velocity
vel =
```

Data Lists

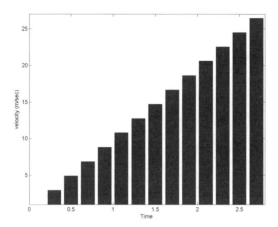

FIGURE 9.7: Bar chart of the velocity of the free-falling object.

```
  Columns 1 through 7
   -0.98    -2.94    -4.90    -6.86    -8.82   -10.78   -12.74
  Columns 8 through 14
  -14.70   -16.66   -18.62   -20.58   -22.54   -24.50   -26.46
Acceleration
acc =
  Columns 1 through 7
   -9.80    -9.80    -9.80    -9.80    -9.80    -9.80    -9.80
  Columns 8 through 13
   -9.80    -9.80    -9.80    -9.80    -9.80    -9.80
```

Figure 9.6 shows the line chart with the values of velocity as they change with time. Figure 9.7 shows a bar chart of the velocity of the object. The acceleration is not shown graphically because it remains constant with value 9.8 (m/sec^2).

Listing 9.5 shows the complete sequence of MATLAB/Octave commands that compute the values of velocity in line 16 and acceleration in line 20 of the problem of the free-falling object. These commands are stored in file rfallobj.m.

Listing 9.5: Command file for computing velocity and acceleration.

```
1 % MATLAB/Octave command file
2 % Compute height, velocity, and acceleration
3 % with rates of change, in the free-falling object
4 % File: rfallobj.m
```

```
 5 y0 = 40.0; % initial position
 6 g = 9.8;
 7 tf = linspace(0.0, 2.8, 15); % values of time
 8 m = length(tf);
 9 dtf = diff(tf);
10 hf = y0 - 0.5 * (g * tf.^2); % computed values of height
11 disp('Height of free-falling object')
12 hf
13 disp('Differences of height')
14 dhf = diff(hf)
15 disp('Velocity')
16 vel = dhf./dtf
17 ddtf = dtf(:, 1:m-2);
18 dvel = diff(vel);
19 disp('Acceleration')
20 acc = dvel./ddtf % acceleration
21 v = [0.0 vel]; % adjust for plot
22 ffobj = plot(tf, v, 'k');
23 set(ffobj, 'LineWidth', 2)
24 box off
25 title('Velocity of Free Falling Object')
26 xlabel('Time')
27 ylabel('Velocity (m/sec)')
28 % store the plot in a figure
29 print -deps rffobj.eps
30 print -dpng rffobj.png
31 print -dtiff rffobj.tif
32 vela = abs(vel); % adjusted to absolute value
33 tfv = tf(:,1: m-1)+dtf/2; % adjusted tf for plot
34 bar(tfv, vela)
35 axis([0.0, 2.8, 0.98, 27.0])
36 title('Velocity of Free Falling Object')
37 xlabel('Time')
38 ylabel('velocity (m/sec)')
39 print -deps r2ffobj.eps
40 print -dpng r2ffobj.png
```

9.7 Area under a Curve

A curve is normally defined by a relationship between variables x and y, and this is expressed as a function, $y = f(x)$. A simple method for approximating the area under a curve between the bounds $x = a$ and $x = b$ is to partition the distance $(b-a)$ into several *trapezoids*, compute the areas of the trapezoids, and add all these areas.

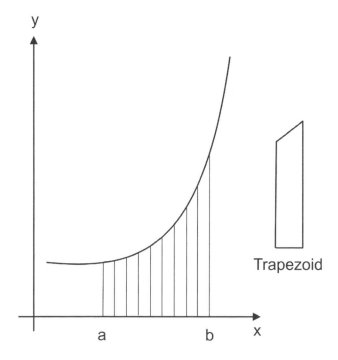

FIGURE 9.8: The area under a curve.

A trapezoid is a four-sided region with two opposite sides parallel. In Figure 9.8, it is the two vertical sides that are parallel. The area of a trapezoid is the average length of the parallel sides, times the distance between them. The area of the trapezoid with width $\Delta x = x_2 - x_1$, is computed as the width, Δx, times the average height, $(y_2 + y_1)/2$.

$$A_t - \Delta x \frac{y_1 + y_2}{2}$$

For the interval of $[a,b]$ on variable x, it is divided into $n-1$ equal segments of length Δx. Any value of y_k is defined as $y_k = f(x_k)$. The trapezoid sum to compute the area under the curve for the interval of $[a,b]$, is defined by the summation of the areas of the individual trapezoids and is expressed as follows.

$$A = \sum_{k=2}^{k=n} [\Delta x \frac{1}{2} (f(x_{k-1}) + f(x_k))]$$

The larger the number of trapezoids, the better the approximation to the area under the curve. The area from a to b, with segments: $a = x_1 < x_2 < \ldots < x_n = b$ is given by the following expression:

$$A = \frac{b-a}{2n}[f(x_1) + 2f(x_2) + \ldots + 2f(x_{n-1}) + f(x_n)]$$

In MATLAB and Octave the area under a curve is approximated by the library function *trapz()*. The following command computes the area under a curve defined by variable y, with x as the independent variable, and using the trapezoid method. The command assigns the value of the area computed to variable A.

```
A = trapz(x, y)
```

The following lines execute command file acffobj.m that computes the area under the curve defined by the height of the free-falling object, in the time interval $[0.5, 2.5]$. The complete time interval is divided into 19 subintervals or trapezoids. The total area is computed twice; the first time using the formula for trapezoids and the second time with the library function *trapz()*. The results have the same value: 54.65.

```
>> acffobj
Values of height
hf =
  Columns 1 through 7
    38.77    37.97    36.97    35.77    34.37    32.77    30.97
  Columns 8 through 14
    28.97    26.77    24.37    21.77    18.97    15.97    12.77
  Column 15
     9.37
Area under curve: 54.65
Area using trapz function: 54.65
```

Listing 9.6 shows the MATLAB/Octave commands that compute the area under the curve and that are stored in command file acffobj.m.

Using a larger number of trapezoids gives a better approximation of the total area under a curve in a given interval $[a,b]$. Partitioning the complete interval into 29 subintervals or trapezoids gives the following results:

```
Area under curve: 54.6756
Area using trapz function: 54.6756
```

Listing 9.6: Command file for computing the area under a curve.

```
1  % MATLAB/Octave command file
2  % Compute the area under the curve of the
3  % height of the free-falling object
4  % File: acffobj.m
5  y0 = 40.0; % initial position
6  g = 9.8;
7  n = 30;
8  tf = linspace(0.5, 2.5, n); % values of time
9  hf = y0 - 0.5 * (g * tf.^2); % computed values of height
10 disp('Values of height')
11 hf
12 a = 0.5; % lower bound
13 b = 2.5; % upper bound
14 A = 0.0; % initial value
15 dt = tf(2) - tf(1); % delta t
16 % n = length(hf); % n-1 = number trapezoids
17 for k = 2:n
18    A = A + dt * 0.5 * (hf(k-1)+hf(k));
19 end
20 moutput('Area under curve: ', A)
21 A2 = trapz(tf, hf);
22 moutput('Area using trapz function: ', A2)
```

9.8 Summary

Arrays are collections for storing a number of values of the same type. Each of these values is known as an element. Once the array has been declared the capacity of the array cannot be changed. To refer to an individual element, an integer value, known as the index, is used to indicate the relative position of the element in the array.

Searching consists of finding an element in the array with a key value. Two important search algorithms are linear search and binary search. With arrays, computing the rate of change and the area under a curve can be approximated, as shown in the problem of the free-falling object.

Key Terms		
creating arrays	accessing elements	array capacity
index	array element	element reference
searching	linear search	binary search
key value	algorithm efficiency	summation
accumulator	rate of change	area under curve

Exercises

Exercise 9.1 Implement a computational model that finds the minimum value element in an array and returns the index value of the element found. Implement with MATLAB or Octave.

Exercise 9.2 Develop a computational model that sorts an array using the Insertion sort technique. This sort algorithm divides the array into two parts. The first is initially empty, it is the part of the array with the elements in order. The second part of the array has the elements in the array that still need to be sorted. The algorithm takes the element from the second part and determines the position for it in the first part. To insert this element in a particular position of the first part, the elements to the right of this position need to be shifted one position to the right. Implement with MATLAB or Octave.

Exercise 9.3 Develop a computational model that computes the standard deviation of values in an array. The standard deviation measures the spread, or dispersion, of the values in the array with respect to the average value. Implement with MATLAB or Octave. The standard deviation of array X with N elements is defined as:

$$std = \sqrt{\frac{sqdif}{N-1}},$$

where

$$sqdif = \sum_{j=0}^{N-1}(X_j - Avg)^2.$$

Exercise 9.4 Develop a computational model that inputs and processes the rainfall data for the last five years. For every year, four quarters of rainfall are provided measured in inches. Hint: use a matrix to store these values. The attributes are: the precipitation (in inches), the year, and the quarter. The program must compute the average, minimum, and maximum rainfall per year and per quarter (for the last five years). Implement with MATLAB or Octave.

Chapter 10

Modules

10.1 Introduction

A general approach for problem solving and algorithmic design is to divide the problem into smaller problems that are easier to solve, and then implement the solution to each of these subproblems. The final solution consists of an assembly of these smaller solutions. The partitioning of a problem into smaller parts is known as *decomposition*. These small parts are known as *modules*, which are much easier to develop and manage.

System design usually emphasizes modular structuring, also called modular decomposition. With this approach, the solution to a problem consists of several smaller solutions corresponding to each of the subproblems. A problem is divided into smaller problems (or subproblems), and a solution is designed for each subproblem. These modules are considered building blocks for constructing larger and more complex algorithms.

10.2 Modular Design

Modular design facilitates algorithmic and computational model development and is particularly important in large and complex problems. *Abstraction* is an important aspect of modularity because each module represents a separate and high-level functionality in the overall design of the algorithm. The algorithm is partitioned into two or more simpler modules. Each module is defined on a single concept (a single functional aspect) and is as independent as possible from other modules.

As mentioned previously, a large and complex algorithm can be broken down into several levels of modules. A *top-down approach* or hierarchical approach is used, by first defining the upper-level modules, then the modules in the next level down, and so on. Each module is defined by two or more sub-

164   *Introduction to Elementary Computational Modeling*

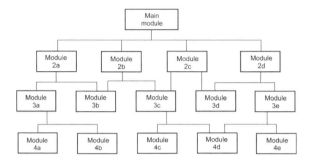

FIGURE 10.1: Modular structure of a computational model.

modules at the next level down. Another way to state this principle is by saying that each module is refined in terms of two or more submodules; this principle is known as *stepwise refinement*.

Figure 10.1 shows a computational model that consists of several modules at various levels of subordination. The main module has four subordinate modules: *2a*, *2b*, *2c*, and *2d*. A module will call or invoke a subordinate module and optionally pass data to it. The main module invokes its subordinate modules in some predefined order. Module *2a* has two subordinate modules: *3a* and *3b*. Some modules may not have subordinate modules.

Modules communicate data via their *interfaces*. Every module has its own interface that specifies how it can be invoked by other modules. An interface indicates the name of the module and the specifications on the type and amount of data that can be transferred to and from the module. Figure 10.2 illustrates the general communication between two modules. Module 2b invokes or calls module 3b. This module communication usually involves transfer of data.

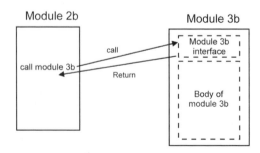

FIGURE 10.2: Module communication and interface.

Modularity and abstraction improve the overall design because a module is much simpler (and smaller) than the overall algorithm; the implementation is also simpler and less prone to errors. The modules have the potential of being reused in other problems or applications. An additional advantage of modularity is that the overall software application is easy to modify and maintain, because the modules are simpler than the overall algorithm, they are more clearly understood, the implementation and testing is easier to manage.

Modules are implemented in various forms depending on programming language used. Some of these forms are: files, libraries, packages, classes, procedures, subroutines, functions, and macros. Software development tools such as MATLAB and Octave provide two basic techniques for implementing modules: scripts and functions. A computational model implemented in MATLAB or Octave can be decomposed into scripts and functions.

10.3 MATLAB and Octave Script Files

A script file is a text file that contains a sequence of Octave/MATLAB commands. The script file is a text file, also known as an m-file, and the file name with an .m extension. To invoke or execute a script, type its name without the .m extension at the Octave or MATLAB command line prompt. Then Octave or MATLAB reads and evaluates the commands in the file. A script can also be invoked from inside another (higher-level) script, in this case, the name of the invoked script file appears as a regular Octave/MATLAB command.

Section 9.5.1.1 (Chapter 9) includes discussion on defining and using a script stored in file aver.m. Listing 10.1 shows the MATLAB and Octave implementation of the algorithm that computes the average value of the elements in an array.

Listing 10.1: MATLAB/Octave command file for computing average.

```
1  % MATLAB/Octave command file to compute average
2  % This script inputs the array size
3  % and the elements of the array from the console
4  % Computes the average value in the array
5  % File: aver.m
6  num = input('Enter array size: ');
7  for j=1:num
8      varr(j) = input('Enter array element: ');
9  end
10 %
11 % Now the average value in array
12 sum = 0.0;
13 for j = 1:num
14     sum = sum + varr(j);
15 end
```

```
16 ave = sum/num;
17 moutput('Average value: ', ave)
```

The following output listing is the result of executing the script *aver* at the MATLAB/Octave prompt.

```
octave-3.2.4.exe:65> aver
Enter array size: 4
Enter array element: 12.4
Enter array element: 2.45
Enter array element: 23.21
Enter array element: 22.1
Average value: 15.04
```

10.4 Functions

A function is the smallest decomposition unit or module in a computational model. A function performs a related sequence of operations to accomplish a particular task. A function has to be defined before it can be used or called. A MATLAB or Octave script can contain one or more function calls.

10.4.1 Function Definition

A function definition includes the interface, local data definitions, and instructions. The interface of a function describes or specifies the name of the function, the type of each parameter needed in calling the function, and the type of data returned by the function. The simplest function definition is one which has an interface that includes only the function name.

The relevant comments for documentation of the function are described in the paragraph that starts with the keyword **description** and ends with '*/'. The name of the function is included after the keyword **function**, and is used when it is called or invoked by some other function. The data variables declared within a function are known only to that function—the scope of the data is *local* to the function. These variables will only exist during execution of the function. The pseudo-code statements for defining a function are:

description
    . . . [comments]
function ⟨ *function_name* ⟩ **is**
    . . . [data definitions]
begin
    . . . [instructions]
endfun ⟨ *function_name* ⟩

The data declarations define local variables and are optional. The rest of the body of a function consists of commands (or instructions) that appear between the keywords **begin** and **endfun**. The following commands in pseudo-code define a simple function for displaying a text message on the screen.

```
description
   displays a message 'Calculating ...'
   on the screen. */
function disp_message is
begin
   display "calculating ..."
endfun disp_message
```

10.4.2 Function Definition in MATLAB and Octave

A function definition in MATLAB and Octave is written starting with the keyword **function**. As with modules of script files, a function definition has to be stored in a file with the name of the function and with an .m extension. The general simple form of a function definition in MATLAB and Octave has the general structure shown as follows:

function ⟨ *function_name* ⟩
    . . . [comments]
    . . . [local data]
    . . . [instructions]

The simple function for displaying a text message on the screen, which was coded previously in pseudo-code, is written in MATLAB and Octave as shown in the following lines of code. This function is stored in file disp_message.m and the commands are shown in Listing 10.2.

Listing 10.2: MATLAB/Octave command file for displaying message.
```
1 function disp_message
2 % MATLAB/Octave command file that defines a function
3 % for displaying a message 'Calculating ...'
4 % File: disp_message.m
5 disp ('Calculating ...');
```

10.4.3 Simple Function Calls

A function *call* is also known as *function invocation* and when a function is called, it starts executing. After the function completes execution, the flow of control is transferred back (returned) to the module or function that issued the call. The module then continues execution from the point after the call. Functions can be reused, which means that a function can be used in another application.

Figure 10.3 illustrates a function call. An upper-level module calls function *B*. After completing its execution, function *B* returns the flow of control to the upper-level module.

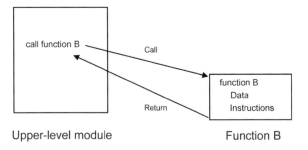

FIGURE 10.3: A simple function call.

In simple function calls, there is no data transfer to or from the function. In pseudo-code, the following statement calls or invokes a function.

call ⟨ `function_name` ⟩

With the following pseudo-code call statement, function *disp_message* is called or invoked from another module or function:

call `disp_message`

In MATLAB and Octave, a function can be invoked or called from a command at the MATLAB/Octave prompt, a script, or from another function. In a command line invoking the function, only the function name is used followed by an empty parenthesis pair. In the following command, function *disp_message* is called.

```
octave-3.2.4.exe:8> disp_message()
Calculating ...
```

10.4.4 Functions with Parameters

Functions that are defined with *parameters* require one or more data items as input values when invoked. The function definition includes the declaration of parameters in the function. For every parameter, its name and type are defined. The general form of a function with parameter declaration in pseudo-code is:

> **description**
>   . . . [documentation]
>   */
> **function** ⟨ *function_name*⟩ **parameters** ⟨ *parameter_list* ⟩
> **is**
>   . . .
> **endfun** ⟨ *function_name* ⟩

The following lines of pseudo-code define function *pdistance* that computes the distance of a point P(x,y) in a Cartesian plane to the origin. The function definition declares two parameters: *x* and *y* and a local variable *dist*.

```
description
  Compute the distance from a point P(x, y)
  to the origin in a Cartesian plane
  */
function pdistance parameters x of type float,
         y of type float
is
variables
   define dist of type float   // local variable
begin
   set dist = sqrt(x^2 + y^2 )
   display "Distance to origin is: ", dist
   return
endfun pdistance
```

The following is the general form of a function definition with parameters in MATLAB and Octave.

> **function** ⟨ *function_name* ⟩ ⟨ *parameter_list* ⟩
> ⟨ *documentation* ⟩
>   . . . [function body]

In MATLAB and Octave, the parameter list consists of the parameters separated by commas and enclosed within parenthesis. Listing 10.3 shows the MATLAB/Octave commands that define function *pdistance*, which computes the distance of a point $P(x,y)$ in a Cartesian plane to the origin. The commands for this function are stored in file `pdistance.m`.

Listing 10.3: Command file for computing distance of a point.

```
1  function pdistance (x, y )
2  % MATLAB/Octave command file to
3  % Compute and display the distance of point P
4  % to the origin in a Cartesian plane
5  % File: pdistance.m
6  % pdistance(x, y)
7     dist = sqrt( x^2 + y^2 );
8     moutput ('Distance to origin is: ', dist)
```

10.4.5 Function Calls with Data

When calling a function, one or more data values can be transferred to the function. Each variable or value passed to the function is known as an *argument* and the complete list of arguments is written in parentheses, with the data items separated by commas. The function can also return data when it completes execution.

In pseudo-code, the argument list appears after the keyword **using** in the **call** statement. The **call** statement for a function call with arguments is:

call ⟨ *function_name* ⟩ **using** ⟨ *argument_list* ⟩

The following example is a call to function *pdistance* with two arguments *a* and *b*. This **call** statement is:

```
call pdistance using a, b
```

A call to function *pdistance*, which was defined previously, requires two arguments. In the following lines of MATLAB/Octave code, function *pdistance* is called with two values: 2.35 and 2.5. When the function executes, it computes and displays the value: 3.4311.

```
octave-3.2.4.exe:24> pdistance(2.35, 2.5)
Distance to origin is: 3.4311
```

10.4.6 Functions with Return Data

A function can transfer data back after it completes execution. Usually, a single value is computed and returned and this can be used in an assignment.

MATLAB and Octave include a library of predefined mathematical functions to compute a wide variety of mathematical computations. The trigonometric functions represent a subgroup of the mathematical functions. For example, the following command in MATLAB/Octave is used to invoke the *sin* function using 1.0 radians as the argument and invoking the same function with 2.25 radians as argument.

```
octave-3.2.4.exe:22> sin(1.0)
ans = 0.8415
octave-3.2.4.exe:23> sin(2.25)
ans =  0.77807
```

The statements in pseudo-code that define the form of a function that returns a value are:

> **description**
>   . . . [documentation]
> */
> **function** ⟨ *function_name* ⟩ **of type** ⟨ *return_type* ⟩ **is**
>   . . .
>   **return** ⟨ *return_value* ⟩
> **endfun** ⟨ *function_name* ⟩

The value to be returned is defined after the **return** keyword. The following example defines a function, *rdistance* that returns the value of the variable *dist*. The pseudo-code for this function definition is:

```
description
  Compute and return the distance from a point P(x, y)
  to the origin in a Cartesian plane
  */
function rdistance return type float
      parameters x of type float, y of type float
is
variables
   define dist of type float    // local variable
begin
   set dist = sqrt(x^2 + y^2 )
   return dist
endfun rdistance
```

The following is the general form of a function definition with return value and parameters in MATLAB/Octave.

function ⟨ *return_data* ⟩ = ⟨ *func_name* ⟩ ⟨ *param_list* ⟩
⟨ *documentation* ⟩
    ... [function body]

Function *rdistance* is defined in MATLAB/Octave command file `rdistance.m` and shown in Listing 10.4; it computes the distance of point P to the origin in a Cartesian plane and returns this value. This function is similar to function *pdistance*.

Listing 10.4: Commands for computing distance from origin to a point.

```
1 function dist = rdistance (x, y )
2 % MATLAB/Octave command file that defines a function
3 % File: rdistance.m
4 % Compute and return  the distance of point P
5 % to the origin in a Cartesian plane
6 % rdistance(x, y)
7   dist = sqrt( x^2 + y^2 );
```

Function *rdistance* can be called directly with two arguments or in an assignment statement, also with two arguments. In the following code, the second call to function *rdistance* assigns the return value of the function to variable *rvar*.

```
octave-3.2.4.exe:26> rdistance(2.35, 2.5)
ans =   3.4311
octave-3.2.4.exe:27> rvar = rdistance(2.35, 2.5)
rvar =   3.4311
```

10.5 Documenting MATLAB and Octave Functions

It is good programming practice to document functions well and follow some well-accepted conventions about what to include in the comments. Some of these conventions are:

- The signature of the function, which includes the name of the function, the input variable(s) and the output variable(s).

- A clear, concise, abstract description of the purpose of the function. Additional comments may be included inside the function that explain the details.

- A description of the input variables.

- A brief description of any preconditions and postconditions.

10.6 Summary

Invoking functions involves several mechanisms for data transfer. Calling simple functions do not involve data transfer between the calling function and the called function. Value-returning functions return a value to the calling function. Calling functions that define one or more parameters involve values sent by the calling function and used as input in the called function.

Key Terms		
methods	functions	method invocation
function call	messages	local data
return value	assignment	parameters
arguments	default values	
top-down approach	stepwise refinement	module interface

Exercises

Exercise 10.1 Why is a function a decomposition unit? Explain.

Exercise 10.2 Why is a script file a decomposition unit? Explain.

Exercise 10.3 Explain the data transfer among functions.

Exercise 10.4 Explain the data transfer among script files.

Exercise 10.5 Explain the level of subordination of functions and script files.

Exercise 10.6 Define a function that includes one or more parameters.

Exercise 10.7 Explain how a function can return more than one value.

Exercise 10.8 What criteria are used to group functions in a file?

Exercise 10.9 Explain the differences between the functions *rdistance* and *pdistance*.

Chapter 11

Mathematical Models: Basic Concepts

11.1 Introduction

A computational model is basically a computer implementation of a mathematical model. Given a real-world problem, we have to define and formulate a mathematical model and then use one or more techniques to implement this model in a computer. A model is a simplified representation of a real-world system and that has a specific purpose. This chapter presents an introduction to simple mathematical models. This includes definitions and explanations of several important concepts related to modeling. Some of these concepts have been briefly introduced in preceding chapters. This chapter also presents an introduction to arithmetic growth to describe the overall behavior of simple mathematical models. The derivation of difference and functional equations is explained; their use in mathematical modeling is illustrated with a few examples. A complete computational model implemented with Octave and MATLAB is presented and discussed.

11.2 From the Real-World to the Abstract World

The general steps in developing a computational model can be summarized as follows:

1. Start with the description of a real-world problem.

2. Understand the problem and derive the specification of the problem. This involves making some simplifying assumptions about the problem.

3. Formulate the problem specification using some form of mathematical relations or graphical techniques.

4. Derive one or more possible solutions to the mathematical formulation.

5. Analyze the mathematical solution and derive a computer implementation.

6. Test the model with real data with real experimental results, if possible.

Step 3 is probably the most important step in modeling. This step involves a transition or conversion from the real-world domain to the *abstract world* domain. The abstract world includes a very simplified view of the original problem. This mapping to the abstract world is illustrated in Figure 11.1.

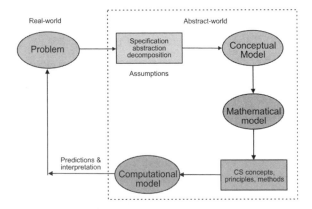

FIGURE 11.1: Real-world to abstract-world mapping.

Step 3 above not only involves a transition to the abstract world from the real world, it also includes the application of mathematical methods that allows us to use or manipulate the data about the problem and help in the study of the original problem.

The three general types of mathematical methods that are used for modeling are:

1. Graphical

2. Numerical

3. Analytical

Graphical methods apply visualization of the data to help understand the data. Various types of graphs can be used, the most common one is the line graph, as shown in Figure 11.2 and Figure 11.3.

Numerical methods directly manipulate the data of the problem to compute various quantities of interest, such as the average change of the population size in a year.

Mathematical Models: Basic Concepts 177

Analytical methods use various forms of relations and equations to allow computation of the various quantities of interest. For example, an equation can be derived that defines how to compute the population size for any given year. Each method has its advantages and limitations. The three methods complement each other and are normally used in modeling.

11.3 Discrete and Continuous Models

In studying how the population changes in a city, we actually need to describe how the population changes with time. In this situation, we have to define a time-dependent model of the population growth in the city specified.

Suppose we have data that was recorded once a year for the last ten years. Therefore, in this model the population changes annually and not instant to instant. In other words, the population changes in jumps at discrete points in time. This type of model is known as a time-dependent *discrete* model.

Figure 11.2 illustrates how the population changes in jumps; it is representing a discrete time-dependent model. Between years 4 and 7 the population changes in small increases. From year 7 to year 10, the population size does not change.

Figure 11.3 shows that the population size changes smoothly with time. There is a value of the population size for every time instant. This represents a *continuous* time-dependent model.

Elementary mathematical models use only simple mathematical notations and techniques to formulate the model. Two types of mathematical equations will be needed to formulate models in this book: *difference* equations and *functional* equations.

11.4 Difference Equations and Data Lists

Data lists were discussed previously and are *collections* of data values. If a list contains the values ordered in some manner, it is known as a *sequence*. Typically, an ordered list is used to represent ordered values of a property of interest in some real problem. Each of these values corresponds to a recorded measure at a specific point in time. In the previous example of population change over a period of five years, the ordered list will contain the value of

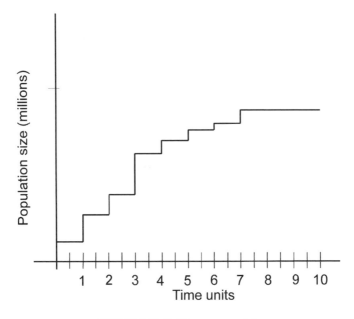

FIGURE 11.2: Discrete model.

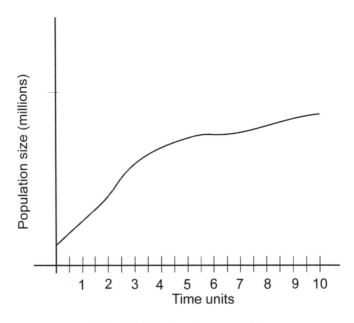

FIGURE 11.3: Continuous model.

the population size for every year. This type of data is discrete data and the expression for the ordered list can be written as:

$$\langle p_1, p_2, p_3, p_4, p_5 \rangle$$

In this expression, p_1 is the value of the population of year 1, p_2 is the value of the population for year 2, p_5 is the value of the population for year 5, and so on. In this case, the length of the list is 5 because it has only five values, or terms.

Another example is the study of changes in electric energy price in a year in Georgia, given the average monthly price. Table 11.1 shows the values of average retail price of electricity for the state of Georgia.[1] The data given corresponds to the price of electric power that has been recorded every month for the last 12 months. This is another example of *discrete data*. The data list is expressed mathematically as:

$$\langle e_1, e_2, e_3, e_4, e_5, e_6, e_7, e_8, e_9, e_{10}, e_{11}, e_{12} \rangle$$

Given the data that has been recorded about a problem and that has been recorded in a list, simple assumptions can be made. One basic assumption is that the quantities in the list increase at a constant rate. This means the increment is assumed to be fixed. If the property is denoted by x, the increment is denoted by Δx, and the value of a term measured at a particular point in time is equal to the value of the preceding term and the increment added to it. This can be expressed as:

$$x_n = x_{n-1} + \Delta x \qquad (11.1)$$

Another assumption is that the values of x are actually always increasing, and not decreasing. This means that the increment is greater than zero, denoted by $\Delta x \geq 0$. At any point in time, the increment of x can be computed as the difference of two consecutive measures of x and has the value given by the expression:

$$\Delta x = x_n - x_{n-1} \qquad (11.2)$$

These last two mathematical expressions, Equation 11.1 and Equation 11.2, are known as *difference equations* and are fundamental for the formulation of simple mathematical models. We can now derive a simple mathematical model for the monthly average price of electric energy, given the collection of monthly recorded energy price in cents per kW-h of the last 12 months. This model is formulated as:

---

[1]U.S. Energy Information Administration - Independent Statistics and Analysis. http://www.eia.gov/

$$e_n = e_{n-1} + \Delta e$$

The initial value of energy price, prior to the first month of consumption, is denoted by e_0, and it normally corresponds to the energy price of a month from the previous year.

TABLE 11.1: Average price of electricity (cents per kW-h) in 2010.

Month	Jan	Feb	Mar	Apr	May	Jun	Jul	Aug
Price	10.22	10.36	10.49	10.60	10.68	10.80	10.88	10.94

Month	Sep	Oct	Nov	Dec
Price	11.05	11.15	11.26	11.40

11.5 Functional Equations

A functional equation has the general form: $y = f(x)$. Where x is the *independent variable* because for every value of x, the function gives a corresponding value for y. In this case, y is a function of x.

An equation that gives the value of x_n at a particular point in time denoted by n, without using the previous value, x_{n-1}, is known as a *functional equation*. In this case the functional equation can also be expressed as: $x = f(n)$. From the data given about a problem and from the difference equation(s), a functional equation can be derived. Using analytical methods, the following mathematical expression can be derived and is an example of a functional equation.

$$x_n = (n-1) \times \Delta x + x_1 \tag{11.3}$$

This equation gives the value of the element x_n as a function of n. In other words, the value x_n can be computed given the value of n. The value of Δx has already been computed. The initial value of variable x is denoted by x_1 and is given in the problem by the first element in the sequence x.

11.6 Validating a Model

Given the data of the problem under study, a model can be derived using difference equations and functional equations. A combination of graphical, numeric, and analytical methods can be applied.

A typical model is used to compute only approximate values for specific properties studied. Validation of a model consists of determining how close the values computed with the model are to the actual values given. For example, starting with the first value of the monthly consumption of electric energy, the model is used to compute the rest of the monthly values of consumption. These values can then be compared to the values given.

If the corresponding values are close enough, the model is considered a reasonable *approximation* to the real system.

11.7 Models with Arithmetic Growth

Arithmetic growth models are the simplest type of mathematical models. Examples of these are time-dependent models, which are models in which a selected property is represented by a variable that changes over time. A simplifying assumption used with these simple models is that the variable increases by equal amounts over equal time intervals. Using x as the variable, the increase is represented by Δx, and the difference equation defined in Equation 11.1 is expressed again as:

$$x_n = x_{n-1} + \Delta x$$

Equation 11.1 and the simplifying assumption of constant growth can be applied to a wide variety of real problems, such as: population growth, monthly price changes of electric energy, yearly oil consumption, and spread of disease.

Using graphical methods, a line or bar chart can be constructed to produce a visual representation of the changes in time of the variable x. Using numerical methods, given the initial value x_0 and once the increase Δx in the property x has been derived, successive values of x can be calculated using Equation 11.1. As mentioned before, with analytical methods, a functional equation can be derived that would allow the direct calculation of variable x_n at any of the n points in time that are included in the data list given. This equation is defined in Equation 11.3 and is expressed again as:

$$x_n = x_1 + \Delta x \times (n-1)$$

11.8 Using MATLAB and Octave to Implement the Model

Using the data of the monthly price of electric energy, the array definitions for the data list given can be implemented in MATLAB/Octave. The array for the monthly price is denoted by e, and the array for months is denoted by m. The data is taken from Table 11.1.

MATLAB and Octave include the *diff(v)* library function that computes the differences of a vector (sequence) of data given. Using the *diff(v)* function, another vector is produced that has the values of the differences of the values in vector v.

```
de = diff(e)
```

11.8.1 MATLAB and Octave Implementation

The code that creates the data vectors and draws the charts is stored in the MATLAB/Octave script file: `priceelect.m`, this is shown in Listing 11.1 and shows the source code for this script. The differences are computed in line 10 of the script file.

In the problem under study, the number of measurements is the total number of months, denoted by n and has a value of 12. To compute the average value of the increments in price of electric energy in the year we can use the general expression for calculating average:

$$\Delta e = \frac{1}{n} \sum_{i=1}^{i=n} de_i$$

This average is implemented in MATLAB and Octave, with the *mean(v)* library (built-in) function that computes the average of the values in a vector v, and using d to denote the average increment Δe. The MATLAB/Octave code is shown in line 12 of the script and is as follows:

```
d = mean(de)
```

The functional equation of the model can be applied to compute the price of electric energy for any month. The new array, c, is created by MATLAB/Octave and contains all the values computed using the functional equation and the average value of the increments. Executing the command file produces the following output:

```
Given data
e =
  Columns 1 through 7
   10.22    10.36    10.49    10.60    10.68    10.80    10.88
  Columns 8 through 12
   10.9400   11.0500   11.1500   11.2600   11.4000
Differences of the given data
de =
  Columns 1 through 7
   0.14     0.13     0.11     0.08     0.12     0.08     0.06
  Columns 8 through 11
   0.1100   0.1000   0.1100   0.1400
Average increment: 0.10727
Calculated data
c =
  Columns 1 through 7
   10.22    10.327   10.43    10.54    10.649   10.75    10.86
  Columns 8 through 12
   10.9709  11.0782  11.1855  11.2927  11.4000
```

11.8.2 Producing the Charts of the Model

Two lists of values for the price of electric energy are available. The first one is given with the problem and is denoted by e, the second list was derived using the functional equation for all 12 months, this list is denoted by c. The list representing the months of the year is denoted by m.

Listing 11.1: Command file for computing the price of electricity.
```
1  % Monthly price for electric energy
2  % File: priceelect.m
3  % Months
4  m = linspace(1,12,12)
5  %Array Monthly price for electric energy
6  disp('Given data')
7  e = [10.22 10.36 10.49 10.60 10.68 10.80 10.88 10.94 11.05
       11.15 11.26 11.40]
8  % differences in sequence e
9  disp('Differences of the given data')
10 de = diff(e)
11 % average of increments
12 d = mean(de);
13 moutput('Average increment: ', d)
14 % Initial value of variable e (before month 1)
15 ei = e(1)-d; % initial value
16 % Creating array of calculated price of electric energy
```

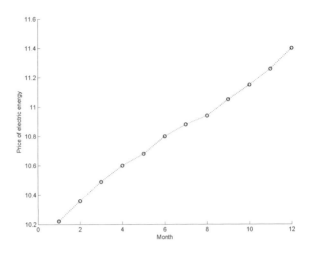

FIGURE 11.4: Monthly price of electric energy.

```
17  disp('Calculated data')
18  c = d*m + ei
19  %%
20  % produce graph of given data
21  pvf = plot(m,e, 'k:o')
22  set(pvf,'LineWidth', 1.5)
23  box off
24  xlabel('Month')
25  ylabel('Price of electric energy')
26  title('Graph of Monthly price of Elecric Energy')
27  print('mpelec.png', '-dpng')
28  print('mpelec.eps', '-deps')
29  % produce graph of given and calculated data
30  pvg = plot(m,e, 'k:o', m, c, 'k-')
31  set(pvg,'LineWidth', 1.5)
32  box off
33  xlabel('Month')
34  ylabel('Price of electric energy')
35  title('Graph of Given and Calulated Monthly Price of
        Electric Energy')
36  legend('Original data', 'Calculated data')
37  print('mpelec2.png', '-dpng')
38  print('mpelec2.eps', '-deps')
```

Figure 11.4 shows the line chart with the original data given in Table 11.1,

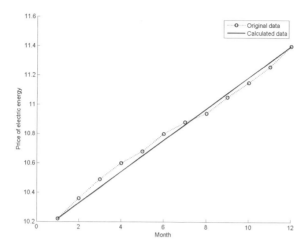

FIGURE 11.5: Monthly price given and calculated of electric energy.

and corresponds to the first chart mentioned. Figure 11.5 shows the combined data of the monthly price given of electric energy and the monthly price calculated using the functional equation.

Using these three data lists, MATLAB and Octave can produce one line chart of array *e* with array *m*, the second chart using array *c* and array *m*, a third chart is produced by combining that data in array *e* and array *c*. The code in the MATLAB/Octave script file: `priceelect.m` creates the arrays and draws the graphs, and is shown in Listing 11.1.

11.9 Summary

A computational model is basically a computer implementation of a mathematical model. This chapter presented some basic concepts of mathematical models. Making simplifying assumptions and using abstraction are important steps in formulating a mathematical model. This involves a transition from the real world to the abstract world. Simple mathematical techniques such as difference equations and functional equations are used. With arithmetic growth models, the values of the differences of the data is constant and the difference equation is linear.

Key Terms		
real world	abstract world	assumptions
specification	mathematical methods	graphical methods
numerical methods	analytical methods	discrete data
data list	ordered list	sequence
discrete models	continuous models	time dependency
difference equations	functional equations	model validation

Exercises

Exercise 11.1 Use a spreadsheet program and produce a line chart of the data list in Table 11.1. Discuss how the price of electric energy changes in a specified period.

Exercise 11.2 Use a spreadsheet program and produce a bar chart of the data list in Table 11.1. Discuss how the price of electric energy changes in a specified period.

Exercise 11.3 Formulate a mathematical model based on a difference equation of the data list in Table 11.1. Use the concepts and principles explained in this chapter.

Exercise 11.4 Discuss the reasons why ordered lists (sequences) of data values are used in the simple mathematical modeling discussed in this chapter.

Exercise 11.5 Using the data in Table 11.1, compute the average increase in price of electric energy per month using Equation 11.1 and/or Equation 11.2. Start with the second month, calculate the price for the rest of the months. Discuss the difference between the data in the table and the corresponding values calculated.

Chapter 12

Models with Quadratic Growth

12.1 Introduction

A model with quadratic growth is one in which the differences in the data are not constant but these differences are growing in some regular manner. Recall that with arithmetic growth the differences are constant. This chapter presents an introduction to mathematical models in which the differences in the data follow a pattern of arithmetic growth.

The computation of difference and functional equations is explained; their use in mathematical modeling is illustrated with a few examples. A complete computational model implemented with Octave and MATLAB is presented and discussed.

12.2 Quadratic Growth

With quadratic growth, the data values in the model do not increase (or decrease) by a constant amount and the differences in the data change linearly.

TABLE 12.1: Number of patients for years 1995–2002.

Year	1995	1996	1997	1998	1999	2000
Patients	5,500	8,500	13,500	20,500	29,500	40,500
Increase	0	3,000	5,000	7,000	9,000	11,000

Year	2001	2002
Patients	53,5000	68,500
Increase	13,000	15,000

187

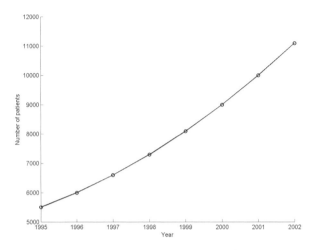

FIGURE 12.1: Number of patients for 1995–2002.

Statistics maintained by a county with several hospitals include the number of patients every year. Table 12.1 shows the data of the number of patients in the years from 1995 to 2002, and their increases. Figure 12.1 shows the graph for the number of patients in the hospital by year from 1995 through 2002. It can be observed that the number of patients from 1995 through 2002 does not follow a straight line, it does not represent a linear relation.

Table 12.1 shows that the number of patients increases every year by a constant number. The differences in the number of patients from year to year increase in a regular pattern, in a linear manner. This implies that the increases of the number of patients follow an arithmetic growth.

12.3 Differences of the Data

The data given in a problem is normally used to set up an ordered list of values, or sequence. This type of list is written as follows:

$$\langle p_1, p_2, p_3, p_4, p_5 \rangle$$

The following example includes the data list given in the problem, these are expressed by the sequence S.

Models with Quadratic Growth

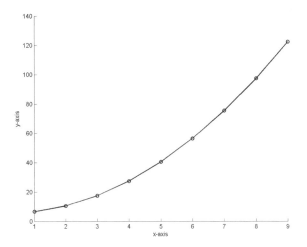

FIGURE 12.2: Original data.

$$S = \langle 6.5, 10.5, 17.5, 27.5, 40.5, 56.5, 75.5, 97.5, 122.5 \rangle$$

Assume that variable x is the independent variable, for example it can be considered the points in time when the values in sequence S were recorded. This is an example of a discrete model. Figure 12.2 shows the graph of the values in sequence S with respect to variable x.

Another sequence, D, is derived that represents the *differences* of the values in sequence S. The values in D are the increases (or decreases) of the values in the first sequence, S.

$$D = \langle 4.0, 7.0, 10.0, 13.0, 16.0, 19.0, 22.0, 25.0 \rangle$$

As mentioned previously, the increases of the values of the differences sequence, D, appear to change linearly and follow an arithmetic growth pattern; this is an important property of quadratic growth models. Figure 12.3 shows the graph with the original data and their differences, with respect to x.

The MATLAB and Octave code that creates the arrays of the independent variable x, the sequence S with the original data, their differences D, their second differences $D2$, and draws the charts is stored in the MATLAB/Octave script file: `differences.m`. When this script executes, the following listing is produced.

```
>> differences
x =
```

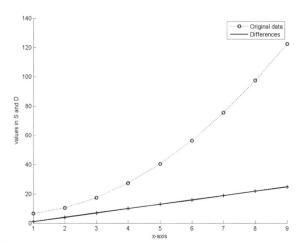

FIGURE 12.3: Original data and differences.

```
         1      2       3       4      5      6      7      8      9
S =
  Columns 1 through 7
    6.50    10.50    17.50    27.50    40.50    56.50    75.50
  Columns 8 through 9
   97.50   122.50
D =
     4      7      10      13      16      19      22      25
D2 =
     3      3       3       3       3       3       3
```

From the listing of executing the script file: differences.m, note that the sequence *D2* contains the values of the *second differences*, that is, the differences of the differences. Sequence *D2* is calculated as the differences of sequence *D*, which contains the differences of the original sequence *S*. In the example, the second differences are constant, all have value 3. This is another important property of quadratic growth models.

As mentioned in previous chapters, MATLAB and Octave include the diff() library function that computes the differences of a sequence of data given. Listing 12.1 shows the source code for this script. The differences are computed in line 9 of the script file.

12.4 Difference Equations

An ordered list or sequence is used to represent ordered values of a property of interest in some real problem. Each of these values corresponds to a recorded measure at a specific point in time.

The expression for the values in a sequence, S, with n values can be written as:

$$S = \langle s_1, s_2, s_3, s_4, s_5 \ldots s_n \rangle$$

In a similar manner, the expression for the values of the differences, D, with m values and with $m = n - 1$, can be written as:

$$D = \langle d_1, d_2, d_3, d_4, d_5 \ldots d_m \rangle$$

Listing 12.1: Command file for computing the differences of a sequence.

```
1  % Plot a sequence and its differences
2  % File: differences.m
3  % x-axis
4  x = linspace(1,9,9)
5  % Array with original sequence
6  S = [6.5 10.5 17.5 27.5 40.5 56.5 75.5 97.5 122.5]
7  % Array with the differences
8  % computed with library function 'diff'
9  D = diff(S)
10 pD = [1.0 D]; % adjusted for plotting
11 % array of differences of the differences
12 D2 = diff(D)
13 %%
14 % produce graph of given data
15 %
16 pvf = plot(x,S, 'k-o')
17 set(pvf,'LineWidth', 1.5)
18 box off
19 xlabel('x-axis')
20 ylabel('y-axis')
21 title('Graph of sequence with original data, S')
22 print('sequenc.png', '-dpng')
23 print('sequenc.eps', '-deps')
24 %
25 % produce graph of given data and differences
26 %
27 pvg = plot(x,S, 'k:o', x, pD, 'k-', x, pD, 'k+')
28 set(pvg,'LineWidth', 1.5)
```

```
29 box off
30 xlabel('x-axis')
31 ylabel('values in S and D')
32 title('Graph of Original Data and Differences')
33 legend('Original data', 'Differences')
34 print('seqdiff.png', '-dpng')
35 print('seqdiff.eps', '-deps')
```

Because the values in the sequence of second differences are all the same, a value is denoted by *dd*.

The value of a term, s_{n+1} in the sequence is equal to the value of the preceding term, s_n and the first term of the differences, d_1, and the single value, *dd*, of the second differences added to it. This equation can be expressed as:

$$s_{n+1} = s_n + d_1 + dd \times (n-1) \tag{12.1}$$

12.5 Functional Equations

Recall that an equation that gives the value of a term x_n without using the previous value, x_{n-1}, is known as a *functional equation*. From the difference equation for quadratic growth models, Equation 12.1, it can be rewritten as:

$$s_n = s_{n-1} + d_1 + dd \times (n-2) \tag{12.2}$$

Equation 12.2 can be manipulated by substituting s_{n-1} for its difference equation, and continuing this procedure until s_1. In this manner, a functional equation can be derived. The following mathematical expression is a general functional equation for quadratic growth models.

$$s_n = s_1 + d_1 \times (n-1) + dd \times (n-2)n/2 \tag{12.3}$$

Equation 12.3 gives the value s_n as a function of n for a quadratic growth model. The value of the first term of the original sequence is denoted by s_1, the value of the first term of the differences is denoted by d_1, and the single value, *dd*, is the second difference that was discussed previously.

12.6 Models with Quadratic Growth

Equation 12.1 and Equation 12.3 represent models with quadratic equations. These can be applied to a wide variety of real problems, such as: computer networks, airline routes, roads and highways, and telephone networks. In these models, the first differences increase in a linear manner (arithmetic growth), as the two examples discussed in previous sections of this chapter. Other models with quadratic growth involve addition of ordered values from several sequences that exhibit arithmetic growth.

12.6.1 Simple Quadratic Growth Models

The following example represents a simple network that connects computers directly to each other. For this a number of links are necessary for the direct connection between computers.

To connect two computers, 1 single link is needed. To connect 3 computers, 3 links are needed. To connect 4 computers, 6 links are needed.

It can be noted that as the number of computers increase, the number of links increase in some pattern. To connect 5 computers, 4 new links are needed to connect the new computer to the 4 computers already connected. This gives a total of 10 links. To connect 6 computers, 5 new links are needed to connect the new computer to the 5 computers that are already connected, a total of 15 links.

Let L_n denote the number of links needed to connect n computers. The difference equation for the number of links can be expressed as:

$$L_n = L_{n-1} + (n-1)$$

This equation has the same form as the general difference equation for quadratic growth, Equation 12.2. The parameters are set as: $d_1 = 0$ and $dd = 1$.

Using the expression for the difference equation for L_n, MATLAB and Octave is used to construct the sequence for L for n varying from 1 to 50. The code is stored in the script file links.m and is shown in Listing 12.2.

Executing the script, all the terms in the sequence L are calculated and a chart is produced. Figure 12.4 shows the graph of the number of links needed to connect n computers.

Listing 12.2: Script to compute the number of links for n computers.
```
1  % This script computes the number of links needed
2  % to connect n computers
3  %
4  m = 50    % limit on number of computers
5  % create arrays for n and L
6  n = linspace(1,m,m)
7  %
8  % Create and initialize array L
9  L = zeros(1, m)
10 % compute the links as n varies from 2 to m
11 for j=2:m
12     L(j) = L(j-1) + (j-1)
13 end
14 %
15 % produce graph of L vs n
16 %
17 pvf = plot(n,L, 'k-o')
18 set(pvf,'LineWidth', 1.5)
19 box off
20 xlabel('Number of computers')
21 ylabel('Number of links')
22 title('Graph of number of links vs n')
23 print('links.png', '-dpng')
24 print('links.eps', '-deps')
```

From the general functional equation, Equation 12.3, the functional equation for the network problem discussed can be expressed as:

$$L_{n+1} = L_1 + (n+1)n/2$$

Notice that L_1 is always zero ($L_1 = 0$) because no link is necessary when there is only one computer ($n = 1$).

$$L_n = n(n-1)/2$$

12.6.2 Models with Sums of Arithmetic Growth

A variety of models have data about a property that follows an arithmetic growth pattern. The summation or running totals of the data of this property is also important. The following example will illustrate this concept.

A county maintains data about the cable installations for multi-purpose services, such as TV, phones, Internet access and others. The data include new cable installations (in thousands) per year and the total number of cable installations per year. This data is shown in Table 12.2.

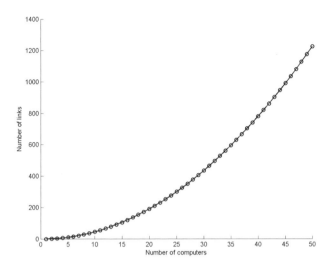

FIGURE 12.4: Number of links to connect n computers.

TABLE 12.2: Number of cable installations for years 1995–2002.

Year	1995	1996	1997	1998	1999	2000	2001	2002
New	1.5	1.9	2.3	2.7	3.1	3.5	3.9	4.3
Sum	1.5	3.4	5.7	8.4	11.5	15.0	18.9	25.2

For this type of data, the general principle is that the new cable installations follow an arithmetic growth pattern, and the summations of the cable installations follow a quadratic growth pattern. Two sequences are needed: a for the new cable installations, and b for the sums of cable installations.

$$a = \langle a_1, a_2, a_3, a_4, a_5 \ldots a_n \rangle \quad b = \langle b_1, b_2, b_3, b_4, b_5 \ldots b_n \rangle$$

To develop a difference equation and a functional equation of the sums of cable installations, a relation of the two sequences, a and b, need to be expressed.

It can be observed from the data in Table 12.2 that the data in sequence a follows an arithmetic growth pattern. The difference equation for a can be expressed as:

$$a_n = a_{n-1} + \Delta a, \quad \Delta a = 400, \quad n = 2 \ldots m$$

The functional equation for a can be expressed as:

$$a_n = 1500 + 400 \times (n-1)$$

It can also be observed from the data in Table 12.2 that the data in sequence b is related to the data in sequence a. The difference equation for sequence b is expressed as:

$$b_n = b_{n-1} + a_n, \quad n = 2 \ldots m$$

Substituting a_n for the expression in its functional equation and using the general functional equation for quadratic growth, Equation 12.3:

$$b_n = b_1 + d_1 \times (n-1) + dd \times (n-2)n/2$$

Finally, the functional equation for the sequence b is expressed as:

$$b_n = 1500 + 1900n + 400(n-1)n/2$$

12.7 Solution and Graphs of Quadratic Equations

A quadratic equation is basically a second degree equation with the general form:

$$y = ax^2 + bx + c, \quad a \neq 0$$

The parameters a, b, and c are known as *coefficients* of the equation. The solution to this type of equation involves computing two roots, x_1 and x_2. These are computed using the following expressions:

$$x_1 = \frac{-b + \sqrt{b^2 - 4ac}}{2a} \quad x_2 = \frac{-b - \sqrt{b^2 - 4ac}}{2a}$$

MATLAB and Octave can be used to solve and to produce graphs of these equations. For the quadratic equation, $2.0x^2 - 4.0x + 1$, the solution computed by MATLAB and Octave is: $x_1 = 1.7071$ and $x_2 = 0.2929$. The graph of this equation is shown in Figure 12.5.

Models with Quadratic Growth

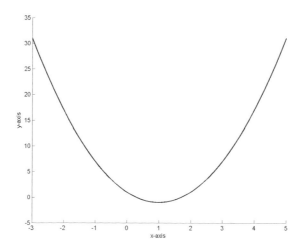

FIGURE 12.5: Graph of a quadratic equation.

Listing 12.3: Commands for computing the roots of a quadratic equation.

```
1  % This script plots a quadratic equation
2  % and computes the roots
3  % File: secdeg2.m
4  m = 50 % number of points to plot
5  a = 2.0   % Coefficients
6  b = -4.0
7  c = 1.0
8  x = linspace(-3.0, 5.0, m)
9  y = zeros(1, m);
10 for j=1:m
11     y(j) = a * x(j)^2 + b*x(j) + c
12 end
13 pvf = plot(x, y, 'k')
14 set(pvf,'LineWidth', 1.5)
15 box off
16 xlabel('x-axis')
17 ylabel('y-axis')
18 title('Graph of Quadratic Equation')
19 print('secdeg2.png', '-dpng')
20 print('secdeg2.eps', '-deps')
21 % solution of the quadratic equation
22 disc = b^2 - 4.0 * a * c
23 x1 = (-b + sqrt(disc))/ (2.0 * a)
24 x2 = (-b - sqrt(disc))/ (2.0 * a)
```

Listing 12.3 contains the MATLAB and Octave code that produces the graph of the equation and computes the solution. This script is stored in file `secdeg2.m`.

12.8 Summary

This chapter presented some basic concepts of quadratic growth in mathematical models. Simple mathematical techniques such as difference equations and functional equations are used in the study of models with quadratic growth. In these models the increases follow an arithmetic growth pattern, the second differences are constants. The most important difference with models with arithmetic growth is that the increases cannot be represented by a straight line, as with arithmetic models. The functional equation of quadratic growth is basically a second degree equation, also known as quadratic equations.

Key Terms		
quadratic growth	nonlinear representation	differences
second differences	network problems	summation
quadratic equation	roots	coefficients

Exercises

Exercise 12.1 In a typical spring day, the temperature varies according to the data recorded in the following table. Formulate the difference equations and functional equations for the temperature. Is this discrete or continuous data? Discuss.

Time	7	8	9	10	11	12	1	2	3	4	5	6
Temp. (F^0)	51	56	60	65	73	78	85	86	84	81	80	70

Models with Quadratic Growth   199

Exercise 12.2   Use a spreadsheet program and produce a line chart and a bar chart of the data list in Table 12.1. Discuss how the number of patients changes in a specified period.

Exercise 12.3   Formulate a mathematical model based on a difference equation of a modified computer network problem similar to one discussed in this chapter. There are three servers and several client computers connected via communication links. All connections between pairs of clients require two links. The connection between servers also require two links. Use the concepts and principles explained in this chapter to derive an equation for the number of links.

Exercise 12.4   Formulate a mathematical model based on a functional equation of a modified computer network problem similar to one discussed in this chapter. There are three servers and several client computers connected via communication links. All connections between pairs of clients require two links. The connection between servers also require two links. Use the concepts and principles explained in this chapter to derive an equation for the number of links.

Exercise 12.5   Formulate a mathematical model based on a functional equation of a modified computer network problem similar to one discussed in this chapter. There are K servers and several client computers connected via communication links. All connections between pairs of clients require two links. The connection of a server to the other $K-1$ servers requires a single link. Use the concepts and principles explained in this chapter to derive an equation for the number of links.

Chapter 13

Models with Polynomial Functions

13.1 Introduction

When using functional equations of higher order than quadratic equations, more general mathematical methods are used. These types of equations are known as polynomial functions. Linear and quadratic equations are special cases of polynomial functions.

This chapter presents several techniques to evaluate and solve polynomial functions. Emphasis is made on graphical and numerical methods. Solutions implemented with Octave and MATLAB are presented and discussed.

13.2 General Forms of Polynomial Functions

As mentioned previously, linear and quadratic equations are special cases of polynomial functions. The degree of a polynomial function is the highest degree among those in its terms. A linear function such as: $y = 3x + 8$, is a polynomial equation of degree 1. A quadratic equation, such as: $y = 4.8x^2 + 3x + 7$, is a polynomial function of degree 2; thus the term second degree equation.

A function such as: $y = 2x^4 + 5x^3 \quad 3x^2 + 7x - 10.5$, is a polynomial function of degree 4 because 4 is the highest exponent of the independent variable x. A polynomial function has the general form:

$$y = p_1 x^n + p_2 x^{n-1} + p_3 x^{n-2} + \ldots p_{k-1} x + p_k$$

This function is a polynomial equation of degree n, and the parameters $p_1, p_2, p_3, \ldots p_k$ are the coefficients of the equation, and are constants.

In addition to the algebraic form of a polynomial function, the graphical form is also important; a polynomial function is represented by a graph.

13.3 Evaluation and Graphs of Polynomial Functions

A polynomial function is evaluated by using various values of the independent variable x and computing the value of the dependent variable y. In a general mathematical sense, a polynomial function defines y as a function of x. With the appropriate expression, several values of polynomial can be computed and graphs can be produced.

13.3.1 Evaluating Polynomial Functions

With computer tools such as MATLAB and Octave, a relatively large number of values of x can be used to evaluate the polynomial function for every value of x. The set of values of x that are used to evaluate a polynomial function are taken from an interval $a \leq x \leq b$ where a is the lower bound of the interval and b is the upper bound. The interval is known as the *domain* of the polynomial function. In a similar manner, the interval of the values of the function y is known as the *range* of the polynomial function.

To use MATLAB and Octave for evaluating a polynomial function $y = f(x)$, two arrays are defined: one with values of x and the other with values of y.

Using the polynomial function $y = 2x^3 - 3x^2 - 36x + 14$ as an example, the two arrays x and y are constructed using MATLAB and Octave. Only 20 values of x are evaluated in this example.

```
x =
  Columns 1 through 7
   -6.00    -5.31    -4.63    -3.94    -3.26    -2.57    -1.89
  Columns 8 through 14
   -1.21    -0.52    0.1589    0.84     1.53     2.21     2.89
  Columns 15 through 20
    3.58     4.26     4.95     5.63     6.31     7.00
y =
  Columns 1 through 7
  -360.00 -229.83  -132.32   -63.65   -19.96    2.58     7.83
  Columns 8 through 14
   -0.36   -18.17   -41.75   -67.25   -90.82  -108.63  -116.83
  Columns 15 through 20
  -111.58  -89.03   -45.34    23.32   120.83   251.00
```

The MATLAB/Octave commands that produce the two arrays x and y, and evaluate the value of the polynomial function for all the values of x, are stored

Models with Polynomial Functions

in the script file `polynom1.m`. Listing 13.1 shows the MATLAB/Octave commands in the script. Lines 5-8 set up the coefficients: a, b, c, and d. Line 10 creates array x with 20 different values from the interval $-6.0 \leq x \leq 7.0$. Lines 12-14 is a loop that computes the value of the function for every value in array x.

The code in script file `polynom1.m` uses a loop to compute the value of a polynomial function for every value of the independent variable x.

Listing 13.1: Script to compute the values of a polynomial function.
```
1  % MATLAB/Octave script to evaluate and plot
      a polynomial equation
2  % File: polynom1.m
3  m = 20; % number of points to plot
4  % Coefficients
5  a = 2.0;
6  b = -3.0;
7  c = -36.0;
8  d = 14.0;
9  %
10 x = linspace(-6.0, 7.0, m)
11 y = zeros(1, m);
12 for j=1:m
13    y(j) = a*x(j)^3 + b*x(j)^2 + c*x(j) + c; % the function
14 end
15 %
16 y
17 pvf = plot(x, y, 'k')
18 set(pvf,'LineWidth', 1.5)
19 box off
20 xlabel('x-axis')
21 ylabel('y-axis')
22 title('Graph of Polynomial Equation')
23 print('polynom1.png', '-dpng')
24 print('polynom1.eps', '-deps')
```

For evaluating polynomial functions, MATLAB and Octave have a library function *polyval()*. A polynomial function is represented in MATLAB and Octave by a vector of *coefficients* in descending order. Function *polyval()* takes a coefficient vector and the vector x as the arguments. The function computes the values of the polynomial (values of y) for all the given values of x.

For example, for the polynomial equation $y = 2x^3 - 3x^2 - 36x + 14$, the coefficient vector is $[2, -3, -36, 14]$. The MATLAB/Octave commands for evaluating the polynomial equation $y = 2x^3 - 3x^2 - 36x + 14$ using *polyval()* are stored in the script file `polynom2.m`; the relevant lines of code are the following:

```
m=20 % number of points
x = linspace(-6.0, 7.0, m)
p = [2 -3 -36 14] % coefficient vector
y = polyval(p, x) % evaluate for all values in x
```

To evaluate the polynomial equation $y = 3x^5 - 2$, the corresponding coefficient vector is $[3, 0, 0, 0, 0, -2]$. The MATLAB and Octave commands use function *polyval()* to evaluate the polynomial in the interval $-2.5 \leq x \leq 2.5$. The commands are stored in the script file polynom3.m and Listing 13.2 shows the MATLAB/Octave commands in the script. Executing this command file produces the following output:

```
>> polynom3
p =
     3     0     0     0     0    -2
x =
  Columns 1 through 7
   -2.50   -2.237   -1.97   -1.71   -1.45   -1.18   -0.92
  Columns 8 through 14
   -0.66   -0.39   -0.13    0.13    0.39    0.66    0.92
  Columns 15 through 20
    1.18    1.45    1.71    1.97    2.23    2.50
y =
  Columns 1 through 7
 -294.97 -169.99  -91.85  -45.93  -21.05   -8.98   -3.99
  Columns 8 through 14
   -2.37   -2.03   -2.0    -1.99   -1.97   -1.63   -0.01
  Columns 15 through 20
    4.98   17.05   41.93   87.85  165.99  290.97
```

Listing 13.2: Script to compute the values of a polynomial function.
```
 1 % MATLAB/Octave script to compute the values and plot
 2 %   a polynomial equation of degree 5
 3 % File: polynom3.m
 4 m = 20; % number of points
 5 % Coefficients
 6 p = [3 0 0 0 0 -2] % coefficient vector
 7 x = linspace(-2.5, 2.5, m) % x vector
 8 y = polyval(p, x) % evaluate for all values in x
 9 %
10 pvf = plot(x, y, 'k')
11 set(pvf,'LineWidth', 1.5)
12 box off
13 xlabel('x-axis')
14 ylabel('y-axis')
15 title('Graph of Polynomial Equation')
```

```
16 print('polynom3.png', '-dpng')
17 print('polynom3.eps', '-deps')
```

13.3.2 Generating Graphs of Polynomial Functions

With the values in arrays of x and y computed, a graph can easily be produced. Figure 13.1 shows the graph of the polynomial function $y = 2x^3 - 3x^2 - 36x + 14$ with the data computed previously. Executing the script files `polynom1.m` and `polynom2.m` uses the *plot()* library function to produce the graph, in addition to evaluating the polynomial function given by the coefficients vector. Lines 17-24 in the `polynom1.m` script file produces the graph shown in Figure 13.1.

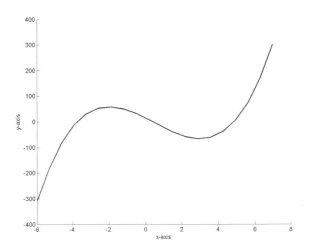

FIGURE 13.1: Graph of the equation $y = 2x^3 - 3x^2 - 36x + 14$.

In a similar manner, executing the command file `polynom3.m` evaluates the polynomial and produces a graph of polynomial equation $y = 3x^5 - 2$. Figure 13.2 shows the graph of this polynomial function.

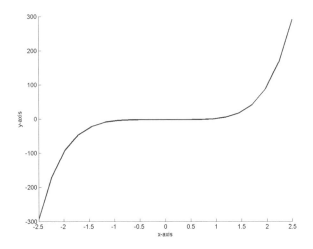

FIGURE 13.2: Graph of the equation $y = 3x^5 - 2$.

13.4 Solution to Polynomial Equations

Recall from Chapter 12 that the solution to a quadratic equation, which is a second degree equation, is relatively straightforward. The solution to this equation involves finding two values of x that gives y value zero. These two values of x are known as the *roots* of the function.

For higher order functional equations, or polynomial functions, the degree of the polynomial determines the number of roots of the function. A polynomial function of degree 7 will have 7 roots.

MATLAB and Octave have the library function *roots()*, that computes the roots of a polynomial function. This function takes the coefficients vector of the polynomial function as argument, and returns a column vector with the roots. This vector is known as the roots vector.

For the polynomial function: $y = 2x^3 - 3x^2 - 36x + 14$, the following MATLAB/Octave lines of code define the coefficients vector and computes the roots by calling function *roots()*. The solution vector, s, has the values of the roots of the polynomial function. The individual values of vector s are: $s_1 = 4.8889$, $s_2 = -3.7688$, and $s_3 = 0.3799$.

```
p = [2 -3 -36 14]  % coefficient vector
p =
        2      -3     -36      14
```

```
s = roots(p) % find the roots of the polynomial
s =
    4.8889
   -3.7688
    0.3799
```

The solution to the polynomial equation $y = 3x^5 - 2$ is computed in the command file `polynom3r.m`. Four of the roots computed are complex values, which are known as *complex roots*. Executing the command file produces the following results:

```
Roots of the polynomial
s =
   -0.7460 + 0.5420i
   -0.7460 - 0.5420i
    0.2849 + 0.8770i
    0.2849 - 0.8770i
    0.9221
```

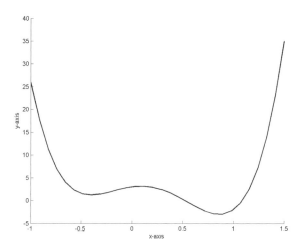

FIGURE 13.3: Graph of equation $y = 23x^4 - 17x^3 - 14x^2 + 3x + 3$.

Script file `polynom4.m` includes the MATLAB/Octave commands that define the coefficients vector, the x vector, evaluates the polynomial computing the y vector, produces the graph, and computes the roots of the polynomial function: $y = 23x^4 - 17x^3 - 14x^2 + 3x + 3$. Figure 13.3 shows the graph of this function for x varying from -1.0 to 1.5. When the script file executes, the following values of the roots are computed:

```
Coefficient vector:  23  -17  -14    3    3
x =
  Columns 1 through 7
   -1.00   -0.91   -0.83   -0.74   -0.65   -0.57   -0.48
  Columns 8 through 14
   -0.39   -0.31   -0.22   -0.14   -0.05    0.03    0.12
  Columns 15 through 21
    0.21    0.29    0.38    0.46    0.55    0.64    0.72
  Columns 22 through 28
    0.81    0.89    0.98    1.07    1.15    1.24    1.32
  Columns 29 through 30
    1.41    1.50
y =
  Columns 1 through 7
   26.00   17.57   11.35    6.95    4.04    2.30    1.45
  Columns 8 through 14
    1.24    1.44    1.87    2.37    2.81    3.08    3.13
  Columns 15 through 21
    2.91    2.42    1.67    0.73   -0.33   -1.39   -2.30
  Columns 22 through 28
   -2.89   -2.95   -2.25   -0.52    2.53    7.25   13.97
  Columns 29 through 30
   23.11   35.06
Roots of the polynomial
s =
    1.087
   -0.436 + 0.194i
   -0.436 - 0.194i
    0.525
```

Listing 13.3 shows the MATLAB/Octave commands in the script polynom4.m.

Listing 13.3: Script to compute the roots of a polynomial function.
```
1  % MATLAB/Octave script that computes the roots and plots
2  %   a polynomial equation
3  % File: polynom4.m
4  m = 30; % number of points
5  % Coefficients
6  p = [23.0 -17.0 -14.0 3.0 3.0];
7  moutput('Coefficient vector: ', p)
8  x = linspace(-1.0, 1.5, m)
9  y = polyval(p, x) % find values of y
10 % find roots of polynomial
11 s = roots(p);
12 disp('Roots of the polynomial')
13 s
```

```
14 pvf = plot(x, y, 'k')
15 set(pvf,'LineWidth', 1.5)
16 box off
17 xlabel('x-axis')
18 ylabel('y-axis')
19 title('Graph of Polynomial Equation')
20 print('polynom4.png', '-dpng')
21 print('polynom4.eps', '-deps')
```

13.5 Summary

This chapter presented some basic techniques for solving polynomial equations that are very useful in computational modeling. Linear and quadratic equations are special cases of polynomial equations. The concepts discussed apply to polynomial functions of any degree. The main emphasis of the chapter are: evaluation of a polynomial function, the graphs of these functions, and solving the functions by computing the roots of the polynomial functions. MATLAB and Octave are used for several case studies.

Key Terms		
polynomial functions	polynomial evaluation	roots
coefficient vector	root vector	evaluation interval
function domain	function range	variable interval

Exercises

Exercise 13.1 Use MATLAB or Octave to evaluate the polynomial function $y = x^4 + 4x^2 + 7$. Find an appropriate interval of x for which the function evaluation is done and plot the graph.

Exercise 13.2 Use MATLAB or Octave to evaluate the polynomial function $y = 3x^5 + 6$. Find an appropriate interval of x for which the function evaluation is done and plot the graph.

Exercise 13.3 Use MATLAB or Octave to evaluate the polynomial function $y = 2x^6 - 1.5x^5 + 5x^4 - 6.5x^3 + 6x^2 - 3x + 4.5$. Find an appropriate interval of x for which the function evaluation is done and plot the graph.

Exercise 13.4 Use MATLAB or Octave to solve the polynomial function $y = x^4 + 4x^2 + 7$.

Exercise 13.5 Use MATLAB or Octave to solve the polynomial function $y = 3x^5 + 6$.

Exercise 13.6 Use MATLAB or Octave to solve the polynomial function $y = 2x^6 - 1.5x^5 + 5x^4 - 6.5x^3 + 6x^2 - 3x + 4.5$.

Chapter 14

Data Estimation and Empirical Modeling

14.1 Introduction

Polynomial functional equations of any order can be evaluated on the various values of the independent variable. If only raw data is available, estimates of the value of the function can be computed for other values of the independent variable, within the bounds of the available set of values of the independent variable. Computing these estimates is carried out using *interpolation* techniques.

If a functional expression is needed that would represent the raw data, then *curve fitting* or *regression* techniques are used. In the general case, the coefficients can be computed (estimated) for a polynomial function of any degree. This technique is often known as empirical modeling. This chapter discusses two general numerical techniques that help estimate data values: interpolation and curve fitting. Solutions implemented with Octave and MATLAB are presented and discussed.

14.2 Interpolation

The given or raw data on a problem usually provides only a limited number of values of data points (x,y). These are normally values of the function y for the corresponding values of variable x. As mentioned previously, using an interpolation technique can estimate intermediate data points. These intermediate values of x and y are not part of the original data. Two well-known interpolation techniques are: linear interpolation and cubic spline interpolation.

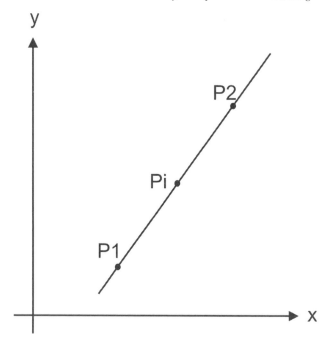

FIGURE 14.1: Graph of linear interpolation of an intermediate data point.

14.2.1 Linear Interpolation

A linear interpolation technique is used with the assumption that an intermediate data point P_i (value of the function y for a value of x) between two known data points: P_1 with coordinates (x_1, y_1) and P_2 with coordinates (x_2, y_2), can be estimated by a straight line between the known data points. In other words, the intermediate data point P_i with coordinates (x_i, y_i) falls on the straight line between the known points: (x_1, y_1) and (x_2, y_2) in the Cartesian plane. Figure 14.1 illustrates the technique of estimating an intermediate data point on a straight line between two given data points.

For example, consider two data points: $(0.5, 1.5)$ and $(6.5, 8.5)$ and an intermediate data point between the given points is to be estimated for $x = 4.25$. Applying a linear interpolation technique, the estimated value computed for y is 5.875. The new intermediate data point is therefore $(4.25, 5.875)$.

MATLAB and Octave have the library function, *interp1()*, that performs linear interpolation given three vectors: x, y, and nx. The first two vectors have the values of the given data points. The third vector, nx, stores the values of x for the new or intermediate data points. The *interp1()* function computes the values for the intermediate data points and stores these in vector ny. In the

Data Estimation and Empirical Modeling

previous example, vector *nx* has only a single value, 4.25. The value of *ny* computed by function *interp1* is: 5.875.

The MATLAB/Octave script that computes and plots an intermediate point at $x = 4.25$ is stored in the file *ltinterpol.m*. The command in line 10 of Listing 14.1 invokes or calls the function *interp1()* with the command ny = interp1(x, y, nx) to compute the value of *ny* for the intermediate data point.

Listing 14.1: Script to compute linear interpolation of data points.

```
1  % MATLAB/Octave script that takes two original data points
2  % and applies linear interpolation to estimate
3  % an intermediate value of y
4  % File: ltinterpol.m
5  % Original data
6  x = [0.5 6.5];
7  y = [1.5 8.5];
8  % new data to be estimated for y
9  nx = 4.25 % intermediate value of x
10 ny = interp1(x,y,nx) % computed estimate of y
11 %
12 pvf = plot(x, y, '-k*', nx, ny, '-ko')
13 set(pvf,'LineWidth', 1.5)
14 box off
15 xlabel('x-axis')
16 ylabel('y-axis')
17 axis([0.0, 7, 0.0, 9.0])
18 title('Graph of estimated intermediate data point')
19 print('ltinterp.png', '-dpng')
20 print('ltinterp.eps', '-deps')
```

The following is a more complete example. There are eight given data points in vectors: *x* and *y*. Vector *x* has equally spaced values of *x* starting at 0 and increasing by 1. Linear interpolation is used to estimate intermediate data points for every value of *x* spaced 0.25, starting at 0 and up to 7. The values to the three given vectors, *x*, *y*, and *nx*, and the value of the computed vector, *ny* are listed as follows:

```
x =
     0     1     2     3     4     5     6     7
y =
     0     3     6     8    12    17    23    26
nx =
  Columns 1 through 7
     0       0.25      0.50      0.75      1.00      1.25      1.50
  Columns 8 through 14
```

```
     1.75      2.00      2.25      2.50      2.75      3.00      3.25
  Columns 15 through 21
     3.50      3.75      4.00      4.25      4.50      4.75      5.00
  Columns 22 through 28
     5.25      5.50      5.75      6.00      6.25      6.50      6.75
  Column 29
     7.00
```

The vector of intermediate points computed by the script is shown as follows:

```
ny =
  Columns 1 through 7
        0      0.75      1.50      2.25      3.00      3.75      4.50
  Columns 8 through 14
     5.25      6.00      6.50      7.00      7.50      8.00      9.00
  Columns 15 through 21
    10.00     11.00     12.00     13.25     14.50     15.75     17.00
  Columns 22 through 28
    18.50     20.00     21.50     23.00     23.75     24.50     25.25
  Column 29
    26.00
```

The estimated values were computed by the MATLAB/Octave script in file *linterpol.m*. The sequence of MATLAB/Octave commands is shown in Listing 14.2. Line 9 in the listing calls function *interp1()* to compute the estimated intermediate points.

Figure 14.2 illustrates the linear interpolation of several intermediate data points given two arrays of the original data points x and y. It can be observed that there are three intermediate points shown as small empty circles, between two original data points, shown as dark small circles.

Listing 14.2: Script to compute linear interpolation of data points.

```
1  % MATLAB/Octave script that takes given data points
2  % and applies linear interpolation to estimate
3  % the function at new values of x
4  % File: linterpol.m
5  x = 0:1:7
6  y = [0 3 6 8 12 17 23 26] % Original data
7  % new data to be estimated for y
8  nx = 0:0.25:7
9  ny = interp1(x,y,nx)
10 pve = plot(x,y,'*-k')
11 set(pve,'LineWidth', 1.5)
12 box off
```

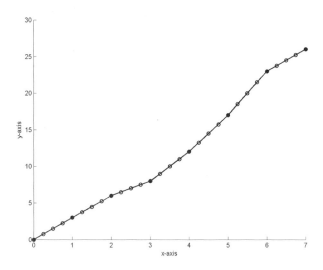

FIGURE 14.2: Graph of linear interpolation of multiple intermediate data points.

```
13 xlabel('x-axis')
14 ylabel('y-axis')
15 title('Graph of original data points')
16 print('lorig.png', '-dpng')
17 print('lorig.eps', '-deps')
18 pvf = plot(x, y, '-k*', nx, ny, '-ko')
19 set(pvf,'LineWidth', 1.5)
20 box off
21 xlabel('x-axis')
22 ylabel('y-axis')
23 title('Graph of estimated data points')
24 print('linterp.png', '-dpng')
25 print('linterp.eps', '-deps')
```

14.2.2 Nonlinear Interpolation

Nonlinear interpolation can generate improved estimates for intermediate data points compared with linear interpolation. MATLAB and Octave include the *cubic spline* interpolation technique. This can be used by calling function *interp1()* with an additional argument that specifies the method used. A smoother curve can be generated using this interpolation technique.

The following example applies the cubic spline interpolation technique to data provided, using an array of values of *nx* for intermediate data points. The MATLAB/Octave script in the file *csinterpol.m* is shown in Listing 14.3. Line 9 calls the function *interp1()* to compute the interpolation using the cubic spline technique.

Executing the script creates the arrays for the original data points in *x* and *y*, and calls the function *interp1()* to compute the estimated intermediate data points in arrays *nx* and *ny*.

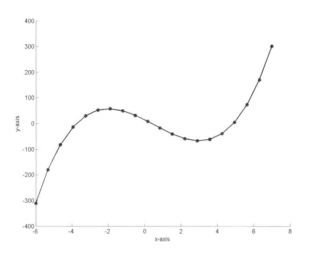

FIGURE 14.3: Graph of given data points.

Figure 14.3 shows the graph of original data points that define a curve. Figure 14.4 shows the graph of the original data points and the estimated data points.

Listing 14.3: Script to compute cubic spline interpolation of data.
```
1  % This script applies cubic spline interpolation and
2  % plots the graphs of original and added estimated data poin
3  % File: csinterpol.m
4  m = 20; % number of points
5  x = linspace(-6.0, 7.0, m) % x vector
6  y = [-310.0 -179.8 -82.3 -13.6 30.0 52.6 57.8 49.6 31.8 8.2
7          -17.2 -40.8 -58.6 -66.8 -61.5 -39.0 4.6 73.3 170.8
            301.0]
8  nx = [-5.0 -4.45 -2.25 -0.25 1.85 3.15 4.75 5.85 6.85]
9  ny = interp1(x, y, nx, 'spline')
10 pvf = plot(x, y, '*-k')
11 set(pvf,'LineWidth', 1.5)
12 box off
```

Data Estimation and Empirical Modeling 217

```
13 xlabel('x-axis')
14 ylabel('y-axis')
15 title('Graph of Given Data')
16 print('polynomcs.png', '-dpng')
17 print('polynomcs.eps', '-deps')
18 pvf = plot(x, y, '*-k', nx, ny, 'ok')
19 set(pvf,'LineWidth', 1.5)
20 box off
21 xlabel('x-axis')
22 ylabel('y-axis')
23 title('Graph of original and Estimated Data')
24 print('polynomcs2.png', '-dpng')
25 print('polynomcs2.eps', '-deps')
```

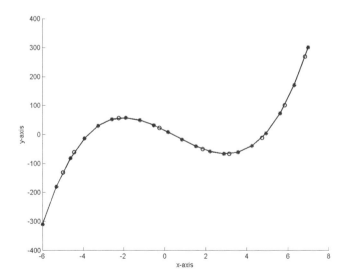

FIGURE 14.4: Graph of given and estimated data points.

14.3 Curve Fitting

Curve fitting techniques, also known as regression techniques, attempt to find the best polynomial expression that represents the given data points. The most widely used curve fitting technique is the *least squares* technique.

Recall that a polynomial function has the general form:

$$y = p_1 x^n + p_2 x^{n-1} + p_3 x^{n-2} + \ldots p_{k-1} x + p_k$$

The parameters $p_1, p_2, p_3, \ldots p_k$ are the coefficients of the equation, which are constants. If a polynomial function of degree 1 is fitted to the given data, the technique is known as *linear regression*, and a straight line is fitted to the data points. As mentioned previously, if the degree of the polynomial is greater than 1, then a curve, instead of a straight line, is fitted to the given data points.

MATLAB and Octave have the *polyfit()* function that computes the coefficients p of the polynomial function of degree n. The three arguments needed to call the function are vectors x and y that define the data points, and the value of the degree of the polynomial, n. The function computes the vector of coefficients of the polynomial function. The MATLAB and Octave command that calls function *polyfit()* is:

```
p = polyfit(x, y , n)
```

Vector p consists of the values of the coefficients of a polynomial function of degree n, given the data points in vectors x and y.

Listing 14.4: Script to compute linear regression of given data points.

```
1  % MATLAB/Octave script that uses given data points
2  % and applies linear REGRESSION to derive
3  % a linear function function
4  % File: linregres.m
5  5 % Original data
6  x = 0:1:7
7  y = [0 3 6 8 12 17 23 26]
8  % coefficients of the polynomial
9  c = polyfit(x, y, 1)
10 nx = linspace(0.0, 7.0, 30); % intermediate data points
11 ny = polyval(c, nx) % evaluate polynomial using nx values
12 pvf = plot(x, y, 'ko', nx, ny, '-k')
13 set(pvf,'LineWidth', 1.5)
14 box off
15 xlabel('x-axis')
16 ylabel('y-axis')
```

```
17 title('Graph of fitted line generated')
18 print('fittlin.png', '-dpng')
19 print('fittlin.eps', '-deps')
```

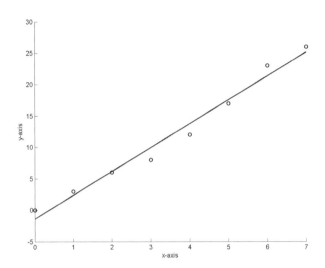

FIGURE 14.5: Graph of fitted linear polynomial.

Listing 14.4 includes the MATLAB/Octave commands that compute the coefficients of a linear polynomial by calling function *polyfit()* in line 9. Function *polyval()* is called in line 11 to evaluate the additional data points in vector *nx*. Finally, the code plots the line and the original data points, which are stored in vectors *x* and *y*. This code is stored in script file linregres.m. Figure 14.5 shows the graph of the fitted line.

Listing 14.5 shows the MATLAB/Octave commands that compute the coefficients of a polynomial of degree 3 by calling function *polyfit()* in line 7. Function *polyval()* is called in line 9 to evaluate the additional data points in vector *nx*. Finally, the code plots the line and the original data points, which are stored in vectors *x* and *y*. This code is stored in script file nl3regres.m.

Figure 14.6 shows the graph of a fitted line to the data points. Figure 14.7 shows the graph of a fitted curve that corresponds to a polynomial of degree 3 to the same data points.

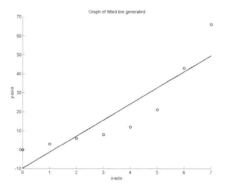

FIGURE 14.6: Regression of linear polynomial.

Listing 14.5: Script to compute nonlinear regression of data points.

```
1  % MATLAB/Octave script that applies non-linear
2  % REGRESSION to derive a polynomial of degree 3
3  % from several given data points
4  % File: nl3regres.m
5  x = 0:1:7
6  y = [0 3 6 8 12 21 43 66] % Original data
7  % coefficients of the polynomial of degree 3
8  c = polyfit(x, y, 3)
9  nx = linspace(0.0, 7.0, 30); % intermediate data points
10 ny = polyval(c, nx) % evaluate polynomial using nx values
11 pvf = plot(x, y, 'ko', nx, ny, '-k')
12 set(pvf,'LineWidth', 1.5)
13 box off
14 xlabel('x-axis')
15 ylabel('y-axis')
16 title('Graph of fitted line generated')
17 print('fittnl3.png', '-dpng')
18 print('fittnl3.eps', '-deps')
```

14.4 Summary

This chapter presented two important techniques that deal with raw data: interpolation and curve fitting. Interpolation is used to compute estimates of the value of the function for intermediate values of the independent variable,

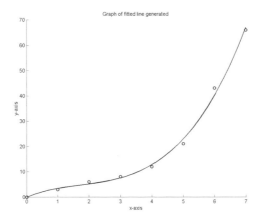

FIGURE 14.7: Regression of a polynomial of degree 3.

within the bounds of the available set of values of the independent variable. Curve fitting or *regression* techniques are used when a polynomial function is needed that would represent the raw data. Solutions implemented with Octave and MATLAB were presented and discussed.

Key Terms

intermediate data	estimated data	raw data
linear interpolation	nonlinear interpolation	curve fitting
cubic spline interpolation	curve fitting	regression

Exercises

Exercise 14.1 On a typical spring day, the temperature (in degrees Fahrenheit) varies according to the data recorded in the following table. Apply linear interpolation to compute estimates of intermediate values of the temperature. Produce a plot of the given and estimated data.

Time	7	8	9	10	11	12	1	2	3	4	5	6
Temp.	51	56	60	65	73	78	85	86	84	81	80	70

Exercise 14.2 On a typical spring day, the temperature varies according to the data recorded in . Apply cubic spline interpolation to compute estimates of intermediate values of the temperature. Produce a plot of the given and estimated data.

Exercise 14.3 On a typical spring day, the temperature varies according to the data recorded in the table of Exercise 14.1. Apply curve fitting to derive the polynomial function of degree 2 that represents the given data. Use the polynomial function to compute estimates of intermediate values of the temperature. Produce a plot of the given and estimated data.

Exercise 14.4 On a typical spring day, the temperature varies according to the data recorded in the table of Exercise 14.1. Apply curve fitting to derive the polynomial function of degree 3 that represents the given data. Use the polynomial function to compute estimates of intermediate values of the temperature. Produce a plot of the given and estimated data.

Exercise 14.5 On a typical spring day, the temperature varies according to the data recorded in the table of Exercise 14.1. Apply curve fitting to derive the polynomial function of degree 4 that represents the given data. Use the polynomial function to compute estimates of intermediate values of the temperature. Produce a plot of the given and estimated data.

Exercise 14.6 Use the data in the following table and apply linear interpolation to compute estimates of intermediate values of the number of patients. Produce a plot of the given and estimated data.

Year	1995	1996	1997	1998	1999	2000
Patients	5,500	8,500	13,500	20,500	29,500	40,500
Increase	0	3,000	5,000	7,000	9,000	11,000

Year	2001	2002
Patients	53,5000	68,500
Increase	13,000	15,000

Exercise 14.7 Use the data in the table of Exercise 14.6 and apply nonlinear interpolation to compute estimates of intermediate values of the number of patients. Produce a plot of the given and estimated data.

Exercise 14.8 Use the data in the table of Exercise 14.6 and apply curve fitting to derive a polynomial function of degree 2. Use this function to compute estimates of intermediate values of the number of patients. Produce a plot of the given and estimated data.

Exercise 14.9 Use the data in the table of Exercise 14.6 and apply curve fitting to derive a polynomial function of degree 3. Use this function to compute estimates of intermediate values of the number of patients. Produce a plot of the given and estimated data.

Chapter 15

Models with Geometric Growth

15.1 Introduction

This chapter presents an introduction to mathematical and computational models in which the data follow a pattern of geometric growth. In such models, the data exhibits growth in a pattern that, in equal intervals of time, the data will increase by an equal percentage or factor.

The difference and functional equations in models with geometric growth are explained; their use in computational modeling is illustrated with a few examples. Several computational models implemented with Octave and MAT-LAB are presented and discussed.

15.2 Basic Concepts of Geometric Growth

As mentioned previously, the data given in a problem is normally used to set up an ordered list of values; this list is known as a sequence. The data in the sequence represents some relevant property of the model and is expressed as a variable s. An individual value of variable s is known as a *term* in the sequence and is denoted by s_n. A sequence with m data values or terms, is written as follows:

$$\langle s_1, s_2, s_3, s_4, s_5, \ldots, s_m \rangle$$

In models with geometric growth, the data increases (or decreases) by an equal percentage or growth factor in equal intervals of time. The difference equation that represents the pattern of geometric growth has the general form:

$$s_{n+1} = c \times s_n \qquad (15.1)$$

In Equation 15.1, the parameter c is constant and represents the *growth*

226 *Introduction to Elementary Computational Modeling*

factor and *n* identifies an individual value such that $n \leq m$. With geometric growth, the data increases or decreases by a fixed factor in equal intervals.

15.2.1 Geometric Growth with Increasing Data

The data in a sequence will successively increase in value when the value of the growth factor is greater than 1.

For example, consider a data sequence that exhibits geometric growth with a growth factor of 1.45 and a starting value of 50.0. The sequence with 8 terms is:

$$\langle 50.0, 72.5, 105.125, 152.43, 221.02, 320.48, 464.70, 673.82 \rangle$$

Figure 15.1 shows a graph of the data with geometric growth. Note that the data increases rapidly.

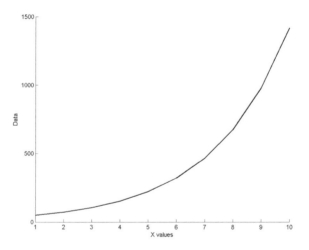

FIGURE 15.1: Data with geometric growth.

15.2.2 Geometric Growth with Decreasing Data

The data in a sequence will successively decrease in value when the value of the growth factor is less than 1.

For example, consider a data sequence that exhibits geometric growth with a growth factor of 0.65 and a starting value of 850.0. The sequence with 10 terms is:

⟨850.0, 552.5, 359.125, 233.43, 151.73, 98.62, 64.10, 41.66, 27.08, 17.60⟩

Figure 15.2 shows a graph of the data with geometric growth. Note that the data decreases rapidly because the growth factor is less than 1.0.

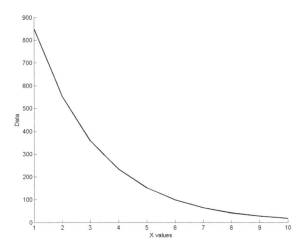

FIGURE 15.2: Data decreasing with geometric growth.

15.2.3   Geometric Growth: Case Study 1

The population of a small town is recorded every year; the increases per year are shown in Table 15.1, which gives the data about the population during the years from 1995 to 2003. The table also shows the population growth factor.

TABLE 15.1: Population of a small town during 1995–2003 (in thousands).

Year	1995	1996	1997	1998	1999	2000	2001	2002	2003
Pop.	81	105	130	162	206	255	288	394	520
Fac.	-	1.29	1.24	1.25	1.27	1.24	1.13	1.36	1.32

Note that although the growth factors are not equal, the data can be considered to grow in a geometric pattern. The values of the growth factor shown in the table are sufficiently close and the average growth factor calculated is 1.263.

The tasks performed developing the computational model are: (1) create the data lists (arrays) of the sequence s with the values of the original data in Table 15.1; (2) compute the average growth factor from the data in the table; (3) compute the values of a second data list y using 1.263 as the average growth factor and the difference equation $y_{n+1} = 1.26 \times y_n$, and (4) draw the graphs.

The MATLAB and Octave commands that perform these tasks are stored in the MATLAB/Octave script file: popstown.m. When this script executes, the following listing is produced.

```
Original data
x =
  Columns 1 through 6
    1995      1996      1997      1998      1999      2000
  Columns 7 through 9
    2001      2002      2003
s =
    81    105    130    162    206    255    288    394    520
m =
     9
Average growth fac: 1.263
Computed data
y =
  Columns 1 through 7
    81.00  102.30  129.21  163.20  206.13  260.35  328.83
  Columns 8 through 9
   415.32  524.56
```

Figure 15.3 shows a graph with two curves; one with the population in the town by year from 1995 through 2003 taken directly from Table 15.1. The other curve shown in the graph of Figure 15.3 is the calculated data applying Equation 15.1 with 1.263 as the growth factor, using the difference equation $s_{n+1} = 1.263 \times s_n$.

Listing 15.1 shows the MATLAB/Octave commands in the script file: popstown.m. The average growth is computed in lines 12-15 of the script file. The computed data is calculated in lines 20-22.

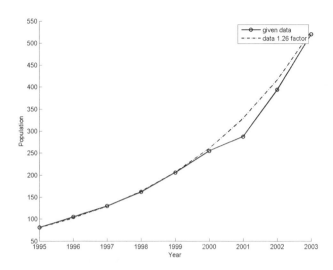

FIGURE 15.3: Population of a small town for 1995–2003.

Listing 15.1: Command file to compute the average growth factor.

```
1  % MATLAB/Octave script that computes average growth factor
2  % From the Population for years 1995-2003
3  % Produces the graphs of the data sequences
4  % File: popstown.m
5  disp('Original data')
6  % x-axis
7  x = [1995 1996 1997 1998 1999 2000 2001 2002 2003]
8  % Array with original sequence
9  s = [81 105 130 162 206 255 288 394 520]
10 m=length(s)   % size of s
11 % compute average growth factor
12 for j=1:m-1
13     f = f+s(j+1)/s(j);
14 end
15 f = f/m;
16 moutput('Average growth fac: ', f);
17 % compute new sequence with aver growth factor
18 y = zeros(1, m);
19 y(1) = 81;
20 for j=1:m-1
21     y(j+1) = f*y(j); % the function
22 end
23 disp('Computed data')
24 y
```

```
25 %
26 % produce graph of given and computed data
27 pvf = plot(x,s, 'k-o', x, y, 'k-.')
28 set(pvf,'LineWidth', 1.5)
29 box off
30 xlabel('Year')
31 ylabel('Population')
32 title('Population of a Small Town 1995--2003')
33 legend('given data', 'data 1.26 factor')
34 print('popstown.png', '-dpng')
35 print('popstown.eps', '-deps')
```

15.2.4 Geometric Growth: Case Study 2

Consider part of a water treatment process in which every application of solvents removes 65% of impurities from the water, to make it more acceptable for human consumption. This treatment has to be performed several times until the water is adequate for human consumption. Assume that when the water has less than 0.6 parts per gallon of impurities, it is adequate for human consumption.

In this problem, the data represents the content of impurities in parts per gallon of water. The initial data is 405 parts per gallon of impurities and the growth factor is 0.35.

The tasks performed in developing the computational model are: (1) compute the values of the data lists or sequence s with the data of the contents of impurities in parts per gallon of water, and (2) plot the graph of the values in sequence s. The MATLAB and Octave commands that implement these tasks are stored in the MATLAB/Octave script file: watertr.m. When this script executes, the following listing is produced.

```
Growth fac: 0.35
Water impurities data
s =
  Columns 1 through 6
    405.0000   141.7500    49.6125    17.3644     6.0775     2.1271
  Columns 7 through 10
     0.7445     0.2606     0.0912     0.0319
```

Figure 15.4 shows the graph of the impurities in parts per gallon of water for several applications of solvents. Listing 15.2 shows the MATLAB commands that compute the impurities in parts per gallon of water, and draws the graphs.

Models with Geometric Growth

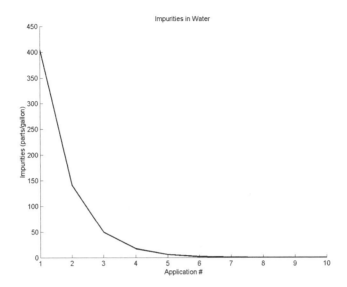

FIGURE 15.4: Impurities in water (parts/gallon).

Listing 15.2: Command file to compute the impurities of water.
```
1  % MATLAB/Octave script
2  % Compute impurities in Water treatment
3  % File: watertr.m
4  % growth factor 0.35
5  m=10
6  % x-axis
7  x = 1:1:m
8  % Array with sequence
9  s = zeros(1, m);
10 m=length(s)   % size of s
11 f = 0.35;
12 moutput('Growth fac: ', f);
13 % compute new sequence with growth factor
14 s(1) = 405;
15 for j=1:m-1
16    s(j+1) = f*s(j); % the function
17 end
18 disp('Water impurities data')
19 s
20 % produce graph of given and computed data
21 pvf = plot(x,s, 'k-')
22 set(pvf,'LineWidth', 1.5)
23 box off
```

```
24 xlabel('Application #')
25 ylabel('Impurities (parts/gallon)')
26 title('Impurities in Water')
27 print('watertr.png', '-dpng')
28 print('watertr.eps', '-deps')
```

15.3 Functional Equations in Geometric Growth

From the difference equation for models with geometric growth, Equation 15.1, the first few terms of sequence s can be written as:

$$s_2 = cs_1$$
$$s_3 = cs_2 = c(cs_1)$$
$$s_4 = cs_3 = c(c(cs_1))$$
$$s_5 = cs_4 = c(c(c(cs_1)))$$
$$s_6 = cs_5 = cc(c(c(cs_1)))$$
$$\ldots$$
$$s_n = c^{n-1}s_1$$

Equation 15.1 was referenced by substituting s_{n-1} for its difference equation, and continuing this procedure up to s_n. In this manner, a functional equation can be derived. Recall that a functional equation gives the value of a term s_n without using the previous value, s_{n-1}. The following mathematical expression is a general functional equation for geometric growth models.

$$s_n = s_1 \times c^{n-1} \tag{15.2}$$

Equation 15.2 gives the value s_n as a function of n for a geometric growth model, with $n \geq 1$. Note that this functional equation includes the fixed value s_1, which is the value of the first term of the data sequence.

A functional equation such as Equation 15.2 is an example of an exponential function because the independent variable, n, is the exponent. This type of growth in the data is also known as *exponential growth*.

Functional equations can be used to answer additional questions about a model. For example: what will the population be 12 years from now? What amount of impurities are left in the water after 8 repetitions of the application of solvents?

When the growth factor does not correspond to the desired unit of time,

Models with Geometric Growth

then instead of n, a more appropriate variable can be used. For example in the first population data in Case Study 1, Section 15.2.3, the variable n represents number of years. To deal with months instead of years, a small substitution in the functional equation is needed. Variable t will represent time, and the starting point of the data is at $t = 0$ with an initial value of y_0. This gives meaning to the concept of a continuous model. Because one year has 12 months and using the same growth factor c as before, the following is a modified functional equation that can be applied when dealing with months.

$$y(t) = y_0 \, c^{t/12} \qquad (15.3)$$

15.4 Properties of Exponential Functions

As mentioned previously, in exponential functions the independent variable is the exponent. In the equation $s_n = s_1 \times c^{n-1}$, the independent variable is n. This is an example of a more general functional equation, $s = f(t)$, in which t is the independent variable.

15.4.1 Exponentiation

A typical exponential equation, such as $s = 95.25 \cdot 1.6^t$, has t as the independent variable and exhibits a graph as shown in Figure 15.5. This equation gives the value of s for any value of t, therefore, the equation represents *continuous data*.

In order to manipulate and solve exponential functional equations, several of its algebraic properties are briefly discussed in this section. The basic rules for exponents can be expressed as follows:

$$\begin{aligned}(y^p)^q &= y^{pq} \\ y^p y^q &= y^{p+q} \\ y^p / y^q &= y^{p-q}\end{aligned}$$

15.4.2 Logarithms

The logarithm is actually an inverse operation to exponentiation. To solve the exponential equation: $y = 10^x$ for x, an equation of the form $x = g(y)$ is needed. The logarithm of y is the power to which 10 would have to be raised to equal y. Solving equation $y = 10^x$, the logarithm is applied to both sides of the equation and this yields: $\log y = x$.

234 Introduction to Elementary Computational Modeling

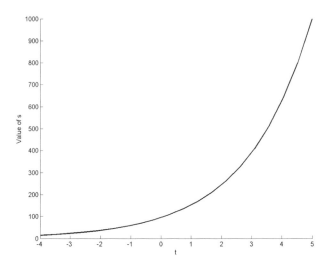

FIGURE 15.5: A typical exponential function, $95.25 \cdot 1.6^t$.

Note that in equation $y = 10^x$, the base of the exponent x is 10. The logarithm base 10 is used to solve the equation, and is denoted by $\log y$. Two other other very common bases are: base 2 and base e.

For an equation such as: $y = 2^x$, the base of the exponent is 2. To solve this equation, logarithm of base 2 is used, and this is denoted as: $\log_2 y$.

The logarithm with base e is known as the *natural logarithm* and the logarithm is of the base e. The value e is an irrational constant approximately equal to 2.718281828. This logarithm is written as $\ln(y)$ or $\log_e y$.

The natural logarithm can be defined for all positive real numbers p as the area under the curve $y = 1/x$ from 1 to p. Logarithms in other bases differ only by a constant multiplier from the natural logarithm, and are usually defined in terms of the latter. Figure 15.6 shows a graph of the values of the natural logarithm for values of variable y between 0 and 3.

Logarithms are used to solve for the half-life, decay constant, or unknown time in exponential decay problems. They are important in many branches of mathematics and the sciences and are used in finance to solve problems involving compound interest.

MATLAB and Octave have three library functions that compute logarithms: *log10()* for base 10, *log()* for base e, and *log2()* for base 2. For example, using MATLAB and Octave to solve $y = 10^x$, for $y = 1.6$:

```
>> y = 1.6
```

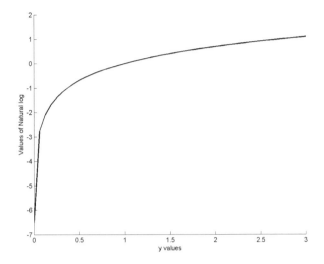

FIGURE 15.6: Natural logarithm.

```
y =
    1.60000000000000
>> x = log10(y)
x =
    0.20411998265592
```

The logarithm base b of y, denoted by $\log_b y$, can be calculated just using logarithm base 10 or with logarithm base e.

$$\log_b y = \frac{\log y}{\log b} = \frac{\log_e y}{\log_e b}$$

Other basic properties of logarithms are:

$$\begin{aligned} \log(xy) &= \log x + \log y \\ \log(x/y) &= \log x - \log y \\ \log(x^p) &= p \log x \end{aligned}$$

Applying logarithms, Equation 15.2 can be solved for n. This can be used to answer additional questions about a model. For example: In what year will the population reach 750,000? How many repetitions of the application of solvents to the water will it take to reach 0.5 parts per gallon of impurities?

15.5 Summary

This chapter presented some basic concepts of geometric growth in mathematical models. The data in these models increase or decrease in a constant growth factor for equal intervals. Simple mathematical techniques such as difference equations and functional equations are used in the study of models with geometric growth. The functional equation of geometric growth is basically an exponential function, so this type of growth is also known as exponential growth. Logarithms are applied to solve exponential equations.

Some important applications involving computational models with geometric growth are: pollution control, human drug treatment, population growth, radioactive decay, and heat transfer.

	Key Terms	
geometric growth	exponential growth	average growth
growth factor	exponent	continuous data
exponentiation rules	logarithms	exponent base

Exercises

Exercise 15.1 In the population problem, Case Study 1 in Section 15.2.3, use the average growth factor already calculated and compute the population up to year 14. For this, modify the corresponding MATLAB/Octave command file. Draw the graphs.

Exercise 15.2 In the population problem, Case Study 1 in Section 15.2.3, use the average growth factor already calculated and compute the population up to year 20. For this, modify the corresponding MATLAB/Octave command file. Draw the graphs.

Exercise 15.3 In the population problem, Case Study 1 in Section 15.2.3, estimate the population for year 18 and month 4. Use the average growth factor already calculated and the modified functional equation, Equation 15.3. Use appropriate MATLAB/Octave commands.

Models with Geometric Growth 237

Exercise 15.4 In the population problem, Case Study 1 in Section 15.2.3, estimate the population for month 50. Use the average growth factor already calculated and the modified functional equation, Equation 15.3. Use appropriate MATLAB/Octave commands.

Exercise 15.5 In the population problem, Case Study 1 in Section 15.2.3, compute the year when the population reaches 750,000. Use the average growth factor and use appropriate MATLAB/Octave commands.

Exercise 15.6 In the population problem, Case Study 1 in Section 15.2.3, compute the month when the population reaches 875,000. Use the average growth factor already calculated and the modified functional equation, Equation 15.3. Use appropriate MATLAB/Octave commands.

Exercise 15.7 In a modification of the water treatment problem, Case Study 2 in Section 15.2.4, an application of solvents removes 57% of impurities in the water. Compute the levels of impurities after several repetitions of the application of solvents. Use this growth factor and use appropriate MATLAB/Octave commands.

Exercise 15.8 In the original water treatment problem, Case Study 2 in Section 15.2.4, compute the levels of impurities after 8 repetitions of the application of solvents. Use this growth factor and use appropriate MATLAB/Octave commands.

Exercise 15.9 In the modified water treatment problem, Exercise 15.8, compute the levels of impurities after 8 repetitions of the application of solvents. Use the growth factor and use appropriate MATLAB/Octave commands.

Exercise 15.10 In the original water treatment problem, Case Study 2 in Section 15.2.4, compute the number of repetitions of the application of solvents that are necessary to reach 0.5 parts per gallon of impurities. Use the growth factor and use appropriate MATLAB/Octave commands.

Chapter 16

Vectors and Matrices

16.1 Introduction

Computations that involve single numbers are known as **scalars**. Data that are collections of data are known as **vectors** and **matrices**. As explained in Chapter 9, vectors are used to implement data lists and sequences.

This chapter presents more detailed explanation of computing with vectors and presents an introduction to matrices.

In general, an array is defined as a data structure that has several values. The values are organized in several ways. The following arrays: X, Y, and Z have their data arranged in different manners. Array X is a one-dimensional array with n elements and it is considered a *row vector* because its elements $x_1, x_2, \ldots, x_n$ are arranged in a single row.

$$X = \begin{bmatrix} x_1 & x_2 & x_3 & \cdots & x_n \end{bmatrix} \qquad Z = \begin{bmatrix} z_1 \\ z_2 \\ z_3 \\ \vdots \\ z_m \end{bmatrix}$$

Array Z is a one-dimensional array; it has m elements organized as a *column vector* because its elements: $z_1, z_2, \ldots, z_m$ are arranged in a single column.

$$Y = \begin{bmatrix} y_{11} & y_{12} & \cdots & y_{1n} \\ y_{21} & y_{22} & \cdots & y_{2n} \\ \vdots & \vdots & \ddots & \vdots \\ y_{m1} & y_{m2} & \cdots & y_{mn} \end{bmatrix}$$

Array Y is a two-dimensional array organized as an $m \times n$ matrix; its elements are arranged in m rows and n columns. The first row of Y consists of elements: $y_{11}, y_{12}, \ldots, y_{1n}$. Its second row consists of elements: $y_{21}, y_{22}, \ldots, y_{2n}$. The last row of Y consists of elements: $y_{m1}, y_{m2}, \ldots, y_{mn}$.

16.2 Vectors

These can be row vectors or column vectors. With MATLAB and Octave, the assignment operator is used in a similar manner as with scalars. The following command applies an assignment after creating a row vector to *P*:

```
P = 1:2:10
P =
    1    3    5    7    9
```

This assignment creates a row vector *P* using 1 as the initial value, 2 as the increment, and 10 as the upper bound or final value. Vector *P* has 5 elements.

$$P = [\, p_1 \; p_2 \; p_3 \; p_4 \; p_5 \,]$$

The value of element p_1 is 1, the value of element p_2 is 3, the value of element p_3 is 5, the value of p_4 is 7, and the value of p_5 is 9.

When specifying the values of a vector, if the number of elements of the vector is too large to fit on a single line, typing a comma and an ellipsis (...) allows continuation on the next line.

A vector can be created by using another vector as part of it. In the following example, vector *B* is created by using vector *P*, which was previously defined. This array operation is also known as *array concatenation*, in which the elements of two or more arrays are joined.

```
>> B = [1 9 3 7 6 P]
B =
  Columns 1 through 9
    1    9    3    7    6    1    3    5    7
  Column 10
    9
```

16.3 Simple Vector Operations

Operations on vectors are performed with a vector and a scalar or with two vectors.

16.3.1 Arithmetic Operations

Adding a scalar to a vector involves adding the scalar value to every element of the vector. For example:

```
>> Q = P + 6.5
Q =
    7.5000    9.5000   11.5000   13.5000   15.5000
```

Adding two vectors involves adding the corresponding elements of each vector and a new vector is created. This add operation on vectors is only possible if the row vectors (or column vectors) are of the same size. In the following example, vectors P and Q are both row vectors of size 5:

```
>> Y = P + Q
Y =
    8.5000   12.5000   16.5000   20.5000   24.5000
```

In a similar manner, subtracting a scalar from a vector can be specified; and subtracting two vectors of the same size.

```
Y2 = P - 2.75
Y2 =
   -1.7500    0.2500    2.2500    4.2500    6.2500
>> Y3 = Y2 + P
Y3 =
   -0.7500    3.2500    7.2500   11.2500   15.2500
```

Multiplying a vector by a scalar results in multiplying each element of the vector by the value of the scalar.

```
A = Q*4.25
A =
   31.8750   40.3750   48.8750   57.3750   65.8750
```

Element by element multiplication is denoted by the operator (.*), also known as dot multiplication, and it multiplies the corresponding elements of the two vectors. This operation can be applied to two row vectors of equal size, or two column vectors of equal size. In the following example, vectors Q and A are dot-multiplied and the results are displayed with a factor of $1.0e+003$, which is actually 10^3.

```
C = Q .* A
C =
   1.0e+003 *
    0.2391    0.3836    0.5621    0.7746    1.0211
```

Element by element division is denoted by the operator (./), also known as dot division, and it divides the corresponding elements of two vectors. This operation can be applied to two row vectors of equal size, or two column vectors of equal size.

```
D = Q ./ A
D =
    0.2353    0.2353    0.2353    0.2353    0.2353
```

The dot powers operation is used to apply element by element powers with the .^ operator. This operation can be applied to a vector and a scalar, or with two row vectors of equal size, or two column vectors of equal size.

```
E = P .^ 3
E =
       1    27   125   343   729

EE = [1 2 3 4 5]
EE =
       1     2     3     4     5
F = P .^ EE
F =
  Columns 1 through 4
       1           9         125        2401
  Column 5
       59049
```

The slicing operation gets part of a vector. This is typically useful when an array needs to be created that is a subarray of another array. The range of index values needs to be specified on the array to be sliced. The range of index values can be specified as an index vector. In the following example, only the elements with index values 1 to 3 are assigned to vector V.

```
V = Q(1:3)
V =
    7.5000    9.5000   11.5000
```

16.3.2 Applying Vector Functions

Various functions can be applied to vectors. Function *size()* gets the number of rows and the number of columns of a vector. For example, vector *B* is a row vector, so the number of rows is 1 and the number of columns is 10.

```
u = size(B)
u =
     1    10
```

Function *length()* returns the largest of the number of rows or the number of columns of the array. The following example calls function *length()*, which returns 5, which is the number of elements of row vector *Q*.

```
length(Q)
ans =
     5
```

Function *max()* gets the maximum value stored in a vector. The following example gets the maximum value in vector *Q*.

```
max(Q)
ans =
    15.5000
```

In addition to the maximum value in a vector, the location of the element with that value may be desired. In the following example the maximum value in vector *Q* and the element location are assigned to another vector with two variables, *u* and *col*.

```
[u col] = max(Q)
u =
    15.5000
col =
     5
```

Function *min()* gets the minimum value stored in a vector. The following example finds the minimum value in vector *Q*.

```
min(Q)
ans =
     7.5000
```

In a similar manner, the minimum value in a vector and its location can be found by using a vector on the left hand of the assignment statement that calls function *min()*.

```
[t col] = min(q)
t =
    7.5000
col =
    1
```

Function *sum* computes the total sum of the values in a vector. The following example computes the summation of the values in vector *P*.

```
s = sum(P)
s =
    25
t = sum(Q)
t =
    57.5000
```

Function *find()* searches an array and returns the index values of the elements for which a relation expression is *True*. In the following example the elements of vector *Q* are examined and the index value of those found to be greater than 10.6 are returned by the function call.

```
>> find(Q > 10.6)
ans =
    3    4    5
```

Function *isequal()* compares two arrays, and if they have the same dimensions and the same elements, returns *True* denoted by 1. If the arrays are not equal, the function returns a 0 (*False*). In the following example, vectors *P* and *Q* are compared by the function and returns 0 because these two vectors are not equal. The vector *Q* is then compared with itself, which obviously is equal and the function now returns 1 (True).

```
isequal(P, Q)
ans =
    0
>> isequal(Q, Q)
ans =
    1
```

16.4 Matrices

A matrix is a two-dimensional array, whereas a vector is a one-dimensional array. MATLAB and Octave process all data as matrices.

A matrix is created by specifying the rows and columns of the array. The semicolon indicates the end of a row. The following example creates a matrix with two rows and three columns:

```
A = [0.5, 2.35,  8.25; 1.8, 7.23, 4.4]
A =
    0.5000    2.3500    8.2500
    1.8000    7.2300    4.4000
size(A)
ans =
    2    3
```

The size of a matrix is specified by the number of rows and columns. In the previous example, function *size()* is invoked to get the number of rows and columns of array A. The function returns a vector with two elements: the first is the number of rows and the second is the number of columns.

A vector is considered a special case of a matrix with one row or one column. A row vector of size n is typically a matrix with one row and n columns. A column vector of size m is a matrix with m rows and one column. Therefore, the operations on vectors also apply to matrices. A *scalar* is considered a matrix with one row and one column.

When specifying the values of a matrix, if the number of elements of the matrix is too large to fit on a single line, typing a comma and an ellipsis (...) allows continuation on the next line.

A matrix can also be created using an existing matrix or existing vectors. This array operation is also known as *array concatenation*, in which the elements of two or more arrays are joined. In the following example, matrix B is created by specifying the values of the first row and the concatenation of matrix A. This builds a matrix with three rows and three columns.

```
B = [ 1 8 12; A]
B =
    1.0000    8.0000   12.0000
    0.5000    2.3500    8.2500
    1.8000    7.2300    4.4000
```

16.4.1 Arithmetic Operations

All the arithmetic operations discussed previously for vectors, also apply to matrices. The following example multiplies the elements of matrix B by the scalar 3.5, and the assignment creates matrix H.

```
H = B * 3.5
H =
     3.5000     28.0000     42.0000
     1.7500      8.2250     28.8750
     6.3000     25.3050     15.4000
```

In the following example, matrix J is created with two rows and three columns. Matrix A and matrix J are added, which means that each element of A is added to the corresponding element of J. This operation is possible because the two matrices have the same number of rows and columns.

```
J = [1 2 3; 4 5 6]
J =
     1     2     3
     4     5     6
K = A + J
K =
     1.5000      4.3500     11.2500
     5.8000     12.2300     10.4000
```

With element-wise multiplication (dot multiplication), each of the elements of a matrix is multiplied by the corresponding element of the second matrix; the results are placed in a new matrix. The operation is specified with the .* operator. For example, matrix A is multiplied with matrix J.

```
L = A .* J
L =
     0.5000      4.7000     24.7500
     7.2000     36.1500     26.4000
```

In matrix multiplication, the number of rows in the first matrix has to equal the number of columns in the second matrix. This is a basic concept of linear algebra.

16.4.2 Function Application

The functions that were discussed for vectors also apply to matrices. The following example calls function *size()* to get the size of matrix L. The function returns two values: the number of rows and the number of columns.

```
size(L)
ans =
     2     3
```

Function *length()* returns the largest of the number of columns or the number of rows. The following example calls the function applied to matrix L and returns 3.

```
length(L)
ans =
     3
```

Function *find()* searches a matrix for those elements that satisfy the specified logical expression. It returns the row and column of every element with which the evaluation of the logical expression is true. In the following example, the function call gets the column and row indexes for every element that satisfies the relational expression. These index values are assigned to a new array that consists of two column vectors: i and j. Only the element at row 1 and column 1; and the element at row 1 and column 2 meet the condition.

```
[i, j] = find(L <= 5.25)
i =
     1
     1
j =
     1
     2
```

16.5 Array Indexing

Individual values or elements of a vector are selected using index values in parentheses. Only one index is needed with a vector because it is a one-dimensional array. With a matrix, two index values are used. The first index value indicates the row number and the second index value the column number.

The following example shows selecting the third element of vector *z*, and multiplying the second element of *z* by 4.5 assigned to variable *w*.

```
z(3)
ans =
    57.5000
w = z(2) * 4.5
w =
    128.2500
```

The index should be an integer value and can be specified as a variable or as an arithmetic expression. The final value of the index should always be within the bounds of the array size. For example:

```
j = 4
z(j)
ans =
    94.5000
w = 3.25 + z(j*2 -3)
w =
    142.7500
```

The length function, which returns the number of elements in a vector, can be used as the upper bound for indexing in a for-loop. The following example accumulates the values of the elements in vector *z*.

```
w = 0.0;
for j=1:length(z)
    w = w + z(j);
end
moutput('Value of w: ', w)
Value of w: 327.5
```

The following example assigns a new value of an element of matrix *L*. The element at row 1 and column 3 is assigned the value 12.75.

```
L(1,3) = 12.75
L =
    0.5000    4.7000   12.7500
    7.2000   36.1500   26.4000
```

The following example extracts the third column of matrix *L* and assigns to *M*.

```
M = L(:,3)
M =
    12.7500
    26.4000
```

The following example extracts columns 1 and 3 of matrix *L* and assigns this to *N*.

```
N = L(:,[1 3])
N =
    0.5000   12.7500
    7.2000   26.4000
```

16.6 Plotting Vectors

As seen in previous chapters, vectors are the most convenient way to store data for plotting. With MATLAB and Octave, calling function *plot()* with a single vector as an argument, plots the index values on the *x*-axis and the element values on the *y*-axis.

Calling function *plot()* with two vectors as arguments, MATLAB and Octave plot the second one as a function of the first; that is, it treats the first vector as a sequence of *x* values and the second as corresponding *y* value and plots a sequence of (x,y) points.

```
X = 1:10
Y = X.^2
plot(X, Y)
```

By default, a blue line is used for plotting. The color, markers, and line styles can be specified by using an addition argument, which is a string.

For example, the calling function *plot()* with string ′ko-′ produces a plot with a black circle at each data point; the hyphen indicates the points should be connected with a line.

```
plot(X, Y, 'ko-')
```

The grid on command includes grid lines to the current plot at the tick marks. By default, MATLAB and Octave start with the `grid off` command. The `box off` command removes the axes box from the current plot.

Labels on the x-axis and y-axis can be included with the `xlabel` command and `ylabel` commands. The `title` command can be used to include a title to the plot. For example:

```
xlabel ('Values of X.')
ylabel ('Values of Y.')
title('Graph of x^2')
```

Figure 16.1 shows a graph produced with the commands discussed in this section.

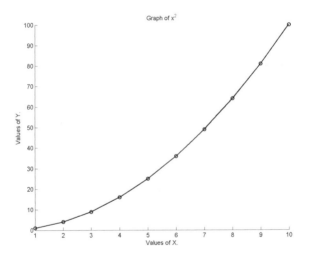

FIGURE 16.1: A simple plot.

16.7 Summary

Computations that involve single numbers are known as **scalars**. Arrays are data structures that store collections of data. To refer to an individual element, an integer value, known as the index, is used to indicate the relative position of the element in the array. MATLAB and Octave manipulate arrays as vectors and matrices. Many operations and functions are defined for vectors and matrices.

Key Terms		
arrays	elements	index
vectors	array elements	matrices
column vector	row vector	double-dimension array
vector operations	vector functions	matrix operations
matrix functions	plotting	

Exercises

Exercise 16.1 Develop a MATLAB/Octave program that computes the values in a vector V sines of the elements of T. The program must assign to T a vector with 75 elements running from 0 to 2π. Plot the elements of V as a function of the elements of T.

Exercise 16.2 Develop a MATLAB/Octave program that finds the index of the first negative number in a vector. If there are no negative numbers, it should set the result to -1.

Exercise 16.3 The Fibonacci series is defined by: $F_n = F_{n-1} + F_{n-2}$. Develop a MATLAB/Octave program that computes a vector with the first n elements of a Fibonacci series. A second vector should also be computed with the ratios of consecutive Fibonacci numbers. The program must plot this second vector.

Exercise 16.4 Develop a MATLAB/Octave program that reads the values of a vector P and computes the element values of vector Q with the cubes of the positive values in vector P. For every element in P that is negative, the corresponding element in Q should be set to zero.

Exercise 16.5 Develop a MATLAB/Octave program that reads the values of a vector P and creates the element values of vector Q with every other element in vector P.

Exercise 16.6 Develop a MATLAB/Octave program that reads the values of a matrix M of m rows and n columns. The program must create a column vector for every column in matrix M, and a row vector for every row in matrix M.

Exercise 16.7 Develop a MATLAB/Octave program that reads the values of a matrix M of m rows and n columns. The program must create a new matrix that has the same number of rows and columns, from the appropriate elements in matrix M. Hint: if $m < n$ then the second matrix would be an $m \times m$ square matrix.

Chapter 17

Text Data

17.1 Introduction

Most of the data defined and manipulated by MATLAB and Octave is numeric in nature. This chapter presents text data also known as *strings*. There is a significant number of string operators and functions that manipulate strings.

17.2 String Vectors

A simple string is basically a sequence of text characters in a row vector, with each character in a column. String column vectors are less common. A character string is denoted by text enclosed in single quotes, and this is typically assigned to a variable. For example:

```
inst = 'Kennesaw State University'
```

17.2.1 String Operations

A string is an array of characters, so each individual character can be referenced using an index value. For example, the eight character—the one with index 8—is accessed by typing 8 in parenthesis. This function applied on string variable *inst*, returns character w.

```
inst(8)
ans =
w
```

The size of the string array can be determined by calling function *size()*. In

253

the following example, MATLAB/Octave return [1 25] that indicates that
the string has one row and 25 columns.

```
size(inst)
ans =
     1    25
```

Using array indexing, a substring can be accessed by specifying the range of index values. In the following example, characters with index values 16 to 25 are retrieved from string *inst*, and assigned to variable *uni*.

```
uni = inst(16:25)
uni =
   University
```

String concatenation joins two strings and creates a longer string. In the following example, the string value of variable *scien* is appended to the string value of variable *inst*, and the new string is assigned to *t*.

```
scien = 'College of Science'
scien =
   College of Science
t = [inst ', ' scien]
t =
   Kennesaw State University, College of Science
```

17.2.2 String Functions

Function *disp()* can be invoked to display the content (or value) of a string variable without including the variable name.

```
disp(t)
Kennesaw State University, College of Science
```

Function *num2str()* can be invoked to convert a numeric value of a variable to a string value. This can be very useful to output data in a program.

```
y = x^2
y =
   20.2500
outvar = ['The value of y is: ' num2str(y) ];
disp (outvar)
   The value of y is: 20.25
```

Text Data

Function *str2num()* can be invoked to convert a string to a numeric value. For example:

```
vt = '9.875'
vt =
    9.875
str2num(vt)
ans =
    9.8750
```

The conversion from string to number is implicitly performed by function *input()*. This function is very useful in scripts that read values from the console. For example:

```
xx = input('Enter value of xx: ')
Enter value of xx: 23.4
xx =
    23.4000
```

The conversion from number to string is implicitly performed by function *disp()*, when including numeric values. For example:

```
disp(xx)
    23.4000
```

Function *ischar()* can be invoked to check if the argument has a string value. If the value of the argument is a string, then the function returns 1 (True). The function returns 0 (False) if the value of the argument is numeric. In the following example, *vt* has a string value and *xx* has a numeric value.

```
ischar(vt)
ans =
    1
ischar(xx)
ans =
    0
```

In addition to lower-case and upper-case letters, a string may include spaces, commas, dots, and the special characters:

```
#, $, %, !, ~, &, -, _, +, =, ?, <, >, /, \, ", ', {, }, (, ),
|, :, ;, @
```

Function *isletter()* examines the characters in the argument string and returns 1 if the character is a letter and returns 0 if it is not. In the following example, there is a space in character of column 5, all the other characters are letters.

```
name = 'John Doe'
name =
    John Doe
isletter(name)
ans =
    1   1   1   1   0   1   1   1
```

Function *isspace()* examines the characters in the argument string and returns 1 if the character is a space and returns 0 if it is not. In the following example, there is a space in character of column 5, all the other characters are letters.

```
isspace(name)
ans =
    0   0   0   0   1   0   0   0
```

Function *lower()* examines the characters in the argument string and converts the letters to lower case. Function *upper()*, converts the characters to upper case.

```
lower(name)
ans =
    john doe
```

Function *findstr()* requires two strings as arguments. It examines the characters in the first string for occurrences of the second string. The function returns the index values where it finds the second string. In the following example, the first call to the function found the substring 'Doe' at column 6 of string *name*. The second call to function *findstr()*, found two occurrences of character 'o' in columns 2 and 7.

```
findstr(name, 'Doe')
ans =
    6
findstr(name, 'o')
ans =
    2   7
```

Function *strrep()* requires three strings as arguments. It examines the characters in the first string for occurrences of the second string. The characters in the second string are replaced with the characters in the third string. In the following example, the call to the function found the substring `'Doe'` and replaces it with the string `'Hunt'` in string *name*.

```
strrep(name, 'Doe', 'Hunt')
ans =
   John Hunt
```

Function *strcat()* requires two strings as arguments. It appends the characters of the second string to the first string. The function returns a longer string. In the following example, the string `', MS'` is concatenated with string *name*.

```
strcat(name, ', MS')
ans =
   John Doe, MS
```

Function *strcmp()* requires two strings as arguments. It compares the characters of the second string with the first string. The function returns a 1 (True) if all the characters are equal, otherwise it returns 0. In the following example, the string 'John Ddd' is compared with string *name*.

```
strcmp(name, 'John Ddd')
ans =
     0
```

17.3 String Matrices

String matrices can be created in a similar manner as numeric matrices. The main restriction is that a string matrix must have the same number of columns on every row. This actually means that the strings on every row must have the same number of characters.

17.4 Summary

Strings are sequences of text characters. They are implemented as row vectors and sometimes as column vectors. To refer to an individual character in a string, an integer value, known as the index, is used to indicate the relative position of the character in the string. MATLAB and Octave manipulate strings using several string operations and functions.

Key Terms		
sstrings	text character	index
row vector	column vector	string matrices
string operations	string functions	substring

Exercises

Exercise 17.1 Reimplement script file *moutput*. Use string function *strcat()*. Test the script with a problem that uses I/O.

Exercise 17.2 Develop a MATLAB/Octave program that takes (as input) a string with a sequence of characters and complement the character sequence into a new string that it returns. This should be performed in such a way that any character ' a ' is converted to a character ' t ' , any character ' g ' is converted to character ' c ' , any character ' t ' to character ' a ' , and any character ' c ' to character ' g ' .

Exercise 17.3 Develop a MATLAB/Octave function that converts the first character in a string to upper case, but only if the first character is a letter.

Exercise 17.4 Develop a MATLAB/Octave function that reverses the characters in a string and returns a new string.

Chapter 18

Advanced Data Structures

18.1 Introduction

Standard arrays are collections of data items of the same type, so they are also known as homogeneous collections. More general data structures are used to define heterogeneous collections of data. These are more restricted in the operations that can be applied with them. To process heterogeneous data collections, an element must first be extracted from the collection, then data operations can be applied.

This chapter presents an introduction to building heterogeneous collections and computing with these data structures.

18.2 Cell Arrays

A cell array is a data collection in which each element is a container or cell with some data item of any type. Each cell of the array is accessed individually to process the data stored in it.

A cell array can be created using two methods: by directly assigning data to individual cells, and by preallocating empty cells given the total number of cells.

An individual cell is accessed using parenthesis or braces. Two general methods for accessing cells are: *cell indexing* and *content addressing*. The following listing uses MATLAB to create simple cell arrays x, Y, and Z using cell indexing. Each cell is enclosed in curly braces.

```
>> Y(1) = {23.5}
Y =
    [23.5000]
>> Z(1) = {34.5}
Z =
```

```
        [34.5000]
>> Z(2) = {'abcde'}
Z =
        [34.5000]       'abcde'
>> X(1) = {[67.5 44.45]}
X =
        {1x1 cell}
>> X(2) = {[12.5 4.55]}
X =
        [1x2 double]    [1x2 double]
```

Using Octave, the previous commands produce the following listing:

```
octave-3.2.4.exe:1> Y(1) = {23.5}
Y =
{
  [1,1] =   23.500
}
octave-3.2.4.exe:2> Z(2) = {'abcde'}
Z =
{
  [1,1] = [](0x0)
  [1,2] = abcde
}
octave-3.2.4.exe:3>  X(1) = {[67.5 44.45]}
X =
{
  [1,1] =
     67.500    44.450
}
octave-3.2.4.exe:4> X(2) = {[12.5 4.55]}
X =
{
  [1,1] =
     67.500    44.450
  [1,2] =
     12.5000    4.5500
}
```

Using content addressing, curly braces are used in the indexing of array cells. The following examples create cell arrays A and B using content addressing.

```
>> A{1} = [12.5 84.25]
A =
```

Advanced Data Structures

```
     [1x2 double]
>> A{2} = 'Science'
A =
     [1x2 double]    'Science'
>> B{1} = [12.5 6.5 2.45; 7.85 14.0 9.0]
B =
     [2x3 double]
>> B{2} = [2.5 41.12]
B =
     [2x3 double]    [1x2 double]
```

A graphical representation of the contents of a cell array is displayed by calling function *cellplot( )*. Figure 18.1 shows the contents of cell array Z.

```
>> cellplot(Z)
```

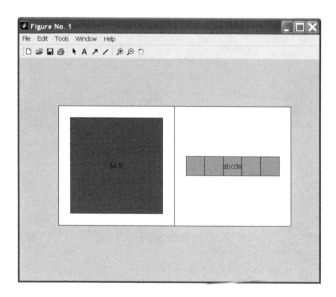

FIGURE 18.1: A simple cell array.

Individual cell contents can be accessed using content addressing. The following example accesses cell 2 of cell array A.

```
>> A{2}
ans =
   Science
```

Function *celldisp()* can be called to display the contents of each cell in a cell array. The following example calls the function to display the cell contents of array cells X and Z.

```
>> celldisp(X)
X{1} =
      67.5000    44.4500
X{2} =
      12.5000    4.5500
>> celldisp(Z)
Z{1} =
      34.5000
Z{2} =
abcde
```

Two-dimensional cell arrays can also be created by assigning values to individual cells.

```
>> Z{1,2} = [4.5 85.2]
Z =
     [34.5000]       [1x2 double]
>> Z{2,2} = ['Data' 'Information']
Z =
     [34.5000]            [1x2 double]
          []         'DataInformation'
>> Z{2,1} = 45.65
Z =
     [34.5000]            [1x2 double]
     [45.6500]       'DataInformation'
```

The following example changes the content of cell {2,2} to an inner cell with two strings.

```
>> Z{2,2} = {['Data'], ['Information']}
Z =
     [34.5000]       [1x2 double]
     [45.6500]       {1x2 cell  }
```

The contents of all cells of cell array Z are displayed by calling function *celldisp()*. The following listing is produced by executing the commands in MATLAB or in Octave.

```
>> celldisp(Z)
Z{1,1} =
   34.5000
Z{2,1} =
   45.6500
Z{1,2} =
   4.5000   85.2000
Z{2,2}{1} =
Data
Z{2,2}{2} =
Information
```

The second method of creating cell arrays is by preallocating empty cells given the number of cells in the array. After creating the cell array, individual cells can be assigned data as shown previously. Function *cell()* is used to preallocate cells to an array. The argument used in the function call is the number of rows and the number of column.

In the following example, cell array *C* is created with 16 cells arranged in 4 rows and 4 columns. Cell array *D* is created with 3 cells arranged in one row.

```
>> C = cell(4)
C =
     []     []     []     []
     []     []     []     []
     []     []     []     []
     []     []     []     []
>> D = cell(1,3)
D =
     []     []     []
```

Using Octave, the following listing is produced by executing the *cell()* function with 4 as the argument.

```
octave-3.2.4.exe:16> C = cell(4)
C =
{
  [1,1] = [](0x0)
  [2,1] = [](0x0)
  [3,1] = [](0x0)
  [4,1] = [](0x0)
  [1,2] = [](0x0)
  [2,2] = [](0x0)
  [3,2] = [](0x0)
  [4,2] = [](0x0)
  [1,3] = [](0x0)
```

```
    [2,3] = [] (0x0)
    [3,3] = [] (0x0)
    [4,3] = [] (0x0)
    [1,4] = [] (0x0)
    [2,4] = [] (0x0)
    [3,4] = [] (0x0)
    [4,4] = [] (0x0)
}
```

Cell arrays can be concatenated just like standard arrays, enclosing them with square brackets. However, all cells on a row in the bracketed expression must have the same number of rows.

```
>> P = [X(1) Y {'KSU'} Z(1,1)]
P =
    [1x2 double]    [23.5000]    'KSU'    [34.5000]
>> celldisp(P)
P{1} =
   67.5000   44.4500
P{2} =
   23.5000
P{3} =
KSU
P{4} =
   34.5000
```

18.3 Structures

More advanced data structures are implemented in MATLAB and Octave with **structures**. Similar to cells, these structures can also store data of any type. The main difference is that structures assign names to the various fields in a cell. This adds a more descriptive structure of heterogeneous arrays.

A data structure is generally known as a *record*, and the various components are known as *fields*. The record and its fields are given appropriate names. The simplest way to build a structure is to assign data values to the various fields of the structure. In the following listing, a record *engine* is created with four fields: *name*, *cost*, *weight*, and *connections*. Specific data values are assigned to these fields.

```
>> engine.name = 'piston';
>> engine.cost = 456.45;
>> engine.weight = 89.5;
>> engine.connections = [211 234 3443 761 289];
>> engine
engine =
         name: 'piston'
         cost: 456.4500
       weight: 89.5000
  connections: [211 234 3443 761 289]
```

The second record of the structure array can be created in a similar manner. The following listing shows the assignments that create the second record of the structure *engine*.

```
>> engine(2).name = 'ring2';
>> engine(2).cost = 56.5;
>> engine(2).weight = 16.5;
>> engine(2).connections = [3211 434];
>> engine
engine =
1x2 struct array with fields:
    name
    cost
    weight
    connections
```

The field names of a structure can be retrieved using function *fieldnames()*. The following example shows a command that invokes this function applied to structure *engine*.

```
>> >> fieldnames(engine)
ans =
    'name'
    'cost'
    'weight'
    'connections'
```

The following example shows how to access the contents of the field *name* of structure *engine*.

```
>> engine.name
ans =
    piston
ans =
    ring2
```

In a similar manner, the following example shows how to access the contents of the field *connections* of structure *engine*.

```
>> engine.connections
ans =
         211         234        3443         761         289
ans =
        3211         434
```

The following example uses direct indexing to access a specific value stored in the second element of field *connections*, in the second record of structure *engine*.

```
>> engine(2).connections(2)
ans =
    434
```

The following example uses direct indexing to access a specific value stored in the third element of field *connections*, in the first record of structure *engine*.

```
>> engine(1).connections(3)
ans =
    3443
```

Using an assignment command a new field can be added to a structure array. The following example adds field *exist* to the first and second records of structure *engine* by assigning 45 and 12.

```
>> engine(1).exist=45
engine =
1x2 struct array with fields:
    name
    cost
    weight
    connections
    exist

>> engine(2).exist=12
engine =
1x2 struct array with fields:
    name
    cost
```

```
    weight
    connections
    exist
>> engine.exist
ans =
    45
ans =
    12
```

A field can be removed from a structure and this is performed by invoking function *rmfield()*. The following example removes field *weight* from structure *engine*.

```
>> rmfield(engine, 'weight')
ans =
1x2 struct array with fields:
    name
    cost
    connections
    exist
```

A MATLAB/Octave function can be defined to create a new record with specified values of the fields in a structure. Assume the general structure as outlined previously for *engine*. Listing 18.1 shows the MATLAB/Octave code of function *nrerec()* that is stored in file newrec.m.

Listing 18.1: MATLAB/Octave function for creating structure record.
```
function ans = newrec(sname, scost, sconn, sexist)
% create a new record of   structure
% MATLAB/Octave function
% File:newrec.m
    ans.name = sname;
    ans.cost = scost;
    ans.connections = sconn;
    ans.exist = sexist;
```

Function *newrec()* is used when creating a new record for the structure *engine*. The following commands show how to create a new record and assign it to element 3 of structure array *engine*.

```
>> engine(3) = newrec('washer2', 34.56, [221 543], 34)
engine =
1x3 struct array with fields:
```

```
        name
        cost
        connections
        exist
>> engine(3)
ans =
            name: 'washer2'
            cost: 34.5600
     connections: [221 543]
           exist: 34
```

18.4 Summary

Heterogeneous collections of data are more powerful for modeling but are more restricted in the operations that can be applied with them. To process heterogeneous data collections, an element must first be extracted from the collection, then data operations can be applied. Cell arrays and structure arrays are explained in this chapter.

Key Terms		
cell arrays	cell elements	index
structure arrays	fields	field names
adding cells	removing cells	cell access
adding fields	removing fields	field access

Exercises

Exercise 18.1 Develop a MATLAB/Octave program that constructs a cell array with data about automobile parts. The data needed are: part number, description, unit cost, sale price, number of units in stock. This array is going to be used for inventory control. Compute the total investment in dollar amount. Find the inventory part with the largest number of units in stock, and the one with the largest cost.

Advanced Data Structures   269

Exercise 18.2 Develop a MATLAB/Octave program that constructs a structure array with data about automobile parts. The data needed are: part number, description, unit cost, sale price, number of units in stock. This array is going to be used for inventory control. Compute the total investment in dollar amount. Find the inventory part with the largest number of units in stock, and the one with the largest cost.

Exercise 18.3 Develop a MATLAB/Octave program that constructs a cell array to maintain data about a CD album collection. The data used are: album title, artist name, producer company, year produced, and a list of songs with title and duration.

Exercise 18.4 Develop a MATLAB/Octave program that constructs a structure array to maintain data about a CD album collection. The data used are: album title, artist name, producer company, year produced, and a list of songs with title and duration.

Exercise 18.5 Develop a MATLAB/Octave program that constructs a cell array to maintain data about course grades in a given course. The data stored are: student name, assignment grades, test grades, and final exam grade. Use the array to compute the average, the maximum and lowest grade in the course.

Exercise 18.6 Develop a MATLAB/Octave program that constructs a structure array to maintain data about course grades in a given course. The data stored are: student name, assignment grades, test grades, and final exam grade. Use the array to compute the average, the maximum and lowest grade in the course.

Exercise 18.7 Develop a MATLAB/Octave program to assign or update the price of used motorcycles. Use a cell array. The data stored is model year, miles, overall state, original price. For every 10,000 miles subtract $2000.00.

Exercise 18.8 Develop a MATLAB/Octave program to assign or update the price of used motorcycles. Use a structure array. The data stored is model year, miles, overall state, original price. For every 10,000 miles subtract $2000.00.

Appendix A

MATLAB and GNU Octave Software

A.1 Introduction

MATLAB® is a registered trademark of The Mathworks, Inc. The following brief documentation is based on that provided by Mathworks. To acquire and get more complete information about the products visit the Mathworks Web page:

http://www.mathworks.com

The MATLAB Environment MATLAB is a high-level technical computing interactive environment and language for numeric computation and data visualization. Additional tasks that can be performed are: algorithm development and data analysis.

MATLAB is used in the following application areas: signal and image processing, communications, control design, test and measurement, financial modeling and analysis, and computational biology.

GNU Octave is also a tool with a high-level language, primarily intended for numerical computations. It provides a convenient command line interface for solving linear and nonlinear problems numerically, and for performing other numerical experiments using a language that is mostly compatible with MATLAB. It may also be used as a batch-oriented language. GNU Octave is freely redistributable software.

A.2 The MATLAB Components

The MATLAB software system consists of the following components:

- *Desktop Tools and Development Environment*: This component includes the set of tools and facilities that help you use and become more productive with MATLAB functions and files. Many of these tools are graphical user interfaces. It includes: the MATLAB desktop and Command Window, an editor and debugger, a code analyzer, and browsers for viewing help, the workspace, and folders.

- *Mathematical Function Library*; This component includes a vast collection of computational algorithms ranging from elementary functions, like sum, sine, cosine, and complex arithmetic, to more sophisticated functions like matrix inverse, matrix eigenvalues, Bessel functions, and fast Fourier transforms.

- *The Language*: This component includes a high-level matrix/array language with control flow statements, functions, data structures, input/output, and object-oriented programming features. It allows "programming in the small" to rapidly create quick programs you do not intend to reuse. You can also do "programming in the large" to create complex application programs intended for reuse.

- *Graphics*: This component includes extensive facilities for displaying vectors and matrices as graphs, as well as annotating and printing these graphs. It includes high-level functions for two-dimensional and three-dimensional data visualization, image processing, animation, and presentation graphics. It also includes low-level functions that allow you to fully customize the appearance of graphics as well as to build complete graphical user interfaces on your MATLAB applications.

- *External Interfaces*: This component includes a library that facilitates writing C/C++ and Fortran programs that interact with MATLAB. It includes facilities for calling routines from MATLAB (dynamic linking), for calling MATLAB as a computational engine, and for reading and writing MAT-files.

A.3 The Desktop

The desktop consists of several tools that are useful for helping to manage the use of the MATLAB software. Figure A.1 shows the default configuration of the MATLAB desktop.

The MATLAB desktop manages several tools and are listed as follows.

- Command History: View a log of or search for the statements you entered in the Command Window, copy them, execute them, and more.

- Command Window: Run MATLAB language statements.

MATLAB and GNU Octave Software

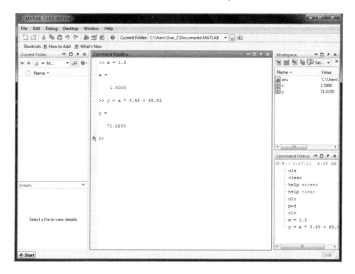

FIGURE A.1: MATLAB desktop.

- Current Folder Browser: View files, perform file operations such as open, find files and file content, and manage and tune your files.

- Editor: Create, edit, debug, and analyze files containing MATLAB language statements.

- File Exchange: Access a repository of files, created by users for sharing with other users, at the MathWorks Web site.

- Figures: Create, modify, view, and print figures generated with MATLAB.

- File and Folder Comparisons: View line-by-line differences between two files.

- Help Browser: View and search the documentation and demos for all MathWorks products.

- Profiler: Improve the performance of your MATLAB code.

- Start Button: Run tools and access documentation for all MathWorks products, and create and use tool bar shortcuts for MATLAB.

- Variable Editor: View array contents in a table format and edit the values.

- Web Browser: View HTML and related files produced by MATLAB.

- Workspace Browser: View and change the contents of the workspace.

A.4 Starting MATLAB

There are two very common ways to start MATLAB on Windows, these are:

- On Windows, click the Start button and select Programs, then MATLAB, then R2010a, then MATLAB R2010a.

- Double-click the MATLAB shortcut on your Windows desktop.

Alternative ways to start MATLAB are:

- Double-click a file with certain file extensions in the Windows Explorer tool. The installer sets up associations between certain file types and MathWorks products during installation. For example, double-clicking a file with a .m extension in the Windows Explorer tool starts MATLAB and opens the file in the MATLAB Editor.

- Open the DOS window, type the cd command to the folder in which you want to start MATLAB and type matlab at the DOS prompt.

After starting MATLAB, the desktop is shown on the screen, see Figure A.1

A.5 Exiting MATLAB

There are several ways to exit MATLAB. To quit MATLAB, perform one of the following:

1. Click the Close box in the MATLAB desktop.

2. Activate the desktop File menu, and select Exit MATLAB.

3. Type quit at the Command Window prompt.

MATLAB responds in the following manner:

1. Prompting the user to confirm quitting, if that preference is specified.

2. Prompting you to save any unsaved files.

3. Running the finish.m script, if it exists in the current folder or on the search path.

A.6 The Command Window

When the Command Window is not open, access it by selecting Command Window from the Desktop menu. All tools open and are minimized in the desktop, except the Command Window, which is maximized.

The Command Window displays the command prompt >>. MATLAB commands or statements are entered at the Command Window prompt. The prompt `EDU>>` indicates that MATLAB Student Version is used. The prompt indicates that MATLAB is ready to accept input from you. This prompt is also known as the command line.

The Command Window is one of the main tools you use to enter data, run MATLAB code, and display results. Figure A.2 shows the command window, the MATLAB command prompt, and a few commands.

FIGURE A.2: MATLAB Command Window.

A.7 Current User Folder

This is the folder that is currently being used. This folder is the one used to access and to store the variables and the files created by the current MATLAB session. Figure A.3 shows part of the MATLAB desktop with the current folder: `c:\computational_mod\work`.

Typically a user will change from the startup folder to this folder to be able to access and store variables and files. The easiest way to set the current folder is by clicking on the Browser for Folder button on the Current Folder box that appears on the Tool bar (in the MATLAB desktop). Another way to change the current folder is to type the `cd` command on the Command Window. For example: `cd c:\comp\var`

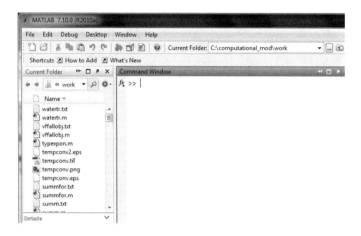

FIGURE A.3: MATLAB Current Folder.

A.8 The Startup Folder

The startup folder is the folder in use when MATLAB starts. The default value for `userpath` is, for example, `Documents/MATLAB` on Microsoft Windows Vista® platforms. A different default value for `userpath` can be specified, or specify a different startup folder. To view the `userpath` value, run the *userpath* function. To specify a location other than the default for `userpath`, or if you do not want to take advantage of `userpath`, make changes with the *userpath* function.

A.9 Using Command Files (Scripts)

A MATLAB command file or script is basically a text file with commands written with the MATLAB language. The simplest way to create a command

MATLAB and GNU Octave Software

file is to start the editor by Selecting File from the desktop menu bar, the selecting New, then selecting Script. This opens the editor, which can be used to write a script file.

Figure A.4 shows the editor with a short script. When the script is completely typed, save with an appropriate name, for example `test1.m`. MATLAB automatically appends the m extension.

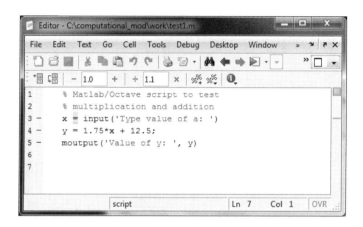

FIGURE A.4: MATLAB Editor.

A command file can be used by typing the name of the script in the command window. For example to execute the script file `test1.m`, type the name at the MATLAB prompt without the m extension. Figure A.5 shows the result of just typing the name of the script at the MATLAB prompt.

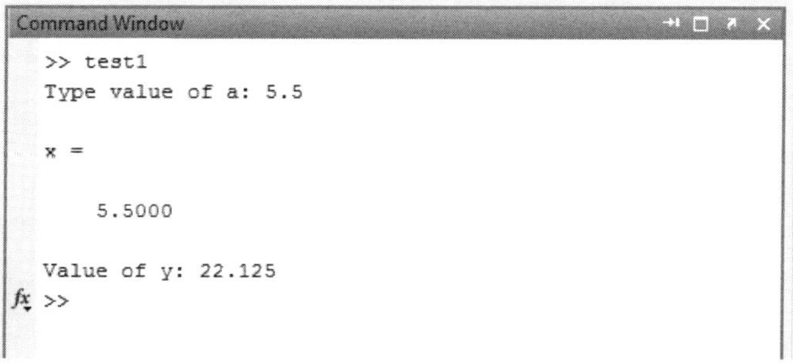

FIGURE A.5: MATLAB Current Folder.

The editor can also be used to modify an existing script. Selecting File from the desktop menu bar, then selecting Open, then selecting the desired script file. This opens the editor with the script.

General Purpose Commands	
`help`	Get online help
`ver`	Get version of software
`who`	Get list of current variables
`whos`	Get more detailed list of current variables
`clc`	Clear the screen (command window)
`clear`	Clear variables and functions in workspace
`load`	Load workspace variables from disk
`save`	Save workspace variables to disk
`quit`	Exit MATLAB

General System Commands	
`path`	Get or set search path
`addpath`	Add directory to search path
`echo`	Echo (show) commands in M-files
`format`	Set output format
`cd`	Change current working directory
`pwd`	Show current working directory
`dir`	List directory
`delete`	Delete file
`!`	Execute operating system command
`profile`	Profile M-file execution time

Control Flow Programming Constructs	
`if`	Condition execution of statements
`else`	Execute if condition is false
`end`	End enclosing statement (if, for, while, switch)
`for`	Repeat statements for a specified number of times
`while`	Repeat statements while condition true
`break`	Terminate enclosing statement (while, for)
`switch`	Switch among several cases on expression evaluation
`case`	Switch statement case
`otherwise`	Default case of switch statement
`return`	Return to invoking function or script

Predefined Variables and Constants	
Name	Return Value
`ans`	Most recent answer (variable). If you do not assign an output variable to an expression, MATLAB automatically stores the result in ans.
`eps`	Floating-point relative accuracy. This is the tolerance the MATLAB software uses in its calculations.
`intmax`	Largest 8-, 16-, 32-, or 64-bit integer the computer can represent.
`intmin`	Smallest 8-, 16-, 32-, or 64-bit integer the computer can represent.
`realmax`	Largest floating-point number the computer can represent.
`realmin`	Smallest positive floating-point number the computer can represent.
`pi`	3.1415926535897...
`i, j`	Imaginary unit.
`inf`	Infinity. Calculations like n/0, where n is any nonzero real value, result in inf.
`NaN`	Not a Number, an invalid numeric value. Expressions like 0/0 and inf/inf result in a NaN, as do arithmetic operations involving a NaN. If n is complex with a zero real part, then $n/0$ returns a value with a NaN real part.
`computer`	Computer type.
`version`	MATLAB version string.

General Mathematical Operators	
Operator	Description
+	Addition
−	Subtraction
.*	Multiplication
./	Right division
.\	Left division
+	Unary plus
−	Unary minus
:	Colon operator
.^	Power
.'	Transpose
'	Complex conjugate transpose
*	Matrix multiplication
/	Matrix right division
\	Matrix left division
^	Matrix power

Relational Operators	
Operator	Description
<	Less than
<=	Less than or equal to
>	Greater than
>=	Greater than or equal to
==	Equal to
~=	Not equal to

Logical Operators	
Operator	Description
&	Returns 1 for every element location that is true (nonzero) in both arrays, and 0 for all other elements.
|	Returns 1 for every element location that is true (nonzero) in either one or the other, or both arrays, and 0 for all other elements.
~	Complements each element of the input array, A.
xor	Returns 1 for every element location that is true (nonzero) in only one array, and 0 for all other elements.

A.10 MATLAB Functions

There is an extensive number and variety of mathematical functions in MATLAB. These functions are grouped into categories of which only the most common and basic are listed here.

Exponential Functions	
exp	Exponential
expm1	Compute exp(x)-1 accurately for small values of x
log	Natural logarithm
log10	Common (base 10) logarithm
log1p	Compute log(1+x) accurately for small values of x
log2	Base 2 logarithm and dissect floating-point numbers into exponent and mantissa
nextpow2	Next higher power of 2
nthroot	Real nth root of real numbers
pow2	Base 2 power and scale floating-point numbers
power	Array power
reallog	Natural logarithm for nonnegative real arrays
realpow	Array power for real-only output
realsqrt	Square root for nonnegative real arrays
sqrt	Square root

\multicolumn{2}{c}{Trigonometric Functions}	
acos	Inverse cosine; in radians
acosd	Inverse cosine; in degrees
acosh	Inverse hyperbolic cosine
acot	Inverse cotangent; in radians
acotd	Inverse cotangent; in degrees
acoth	Inverse hyperbolic cotangent
acsc	Inverse cosecant; in radians
acscd	Inverse cosecant; in degrees
acsch	Inverse hyperbolic cosecant
asec	Inverse secant; in radians
asecd	Inverse secant; in degrees
asech	Inverse hyperbolic secant
asin	Inverse sine; in radians
asind	Inverse sine; in degrees
asinh	Inverse hyperbolic sine
atan	Inverse tangent; in radians
atan2	Four-quadrant inverse tangent
atand	Inverse tangent; result in degrees
atanh	Inverse hyperbolic tangent
cos	Cosine of argument in radians
cosd	Cosine of argument in degrees
cosh	Hyperbolic cosine
cot	Cotangent of argument in radians
cotd	Cotangent of argument in degrees
coth	Hyperbolic cotangent
csc	Cosecant of argument in radians
cscd	Cosecant of argument in degrees
csch	Hyperbolic cosecant
hypot	Square root of sum of squares
sec	Secant of argument in radians
secd	Secant of argument in degrees
sech	Hyperbolic secant
sin	Sine of argument in radians
sind	Sine of argument in degrees
sinh	Hyperbolic sine of argument in radians
tan	Tangent of argument in radians
tand	Tangent of argument in degrees
tanh	Hyperbolic tangent

Date and Time Functions

Date and Time	Output	Function
Current date and time	Date number	`now`
	Date vector	`clock`
	Date string	`datestr(now)`
Current date	Date number	`datenum(date)`
	Date vector	`datevec(date)`
	Date string	`date`
Day of week	Day name or number (1-7)	`weekday`
Last day of month(s)	Vector of one or more days	`eomday`
Date with modified field	Date number	`addtodate`
Calendar for month	6-by-7 matrix of days	`calendar`

Complex Functions

abs	Absolute value and complex magnitude
angle	Phase angle
complex	Construct complex data from real and imaginary components
conj	Complex conjugate
cplxpair	Sort complex numbers into complex conjugate pairs
i	Imaginary unit
imag	Imaginary part of complex number
isreal	Check if input is real array
j	Imaginary unit
real	Real part of complex number
sign	Signum function
unwrap	Correct phase angles to produce smoother phase plots

Rounding and Remainder

ceil	Round toward positive infinity
fix	Round toward zero
floor	Round toward negative infinity
idivide	Integer division with rounding option
mod	Modulus after division
rem	Remainder after division
round	Round to nearest integer

Cartesian Coordinate System Conversion

cart2pol	Transform Cartesian coordinates to polar or cylindrical
cart2sph	Transform Cartesian coordinates to spherical
pol2cart	Transform polar or cylindrical coordinates to Cartesian
sph2cart	Transform spherical coordinates to Cartesian

Polynomials	
conv	Convolution and polynomial multiplication
deconv	Deconvolution and polynomial division
poly	Polynomial with specified roots
polyder	Polynomial derivative
polyeig	Polynomial eigenvalue problem
polyfit	Polynomial curve fitting
polyint	Integrate polynomial analytically
polyval	Polynomial evaluation
polyvalm	Matrix polynomial evaluation
residue	Convert between partial fraction expansion and polynomial coefficients
roots	Polynomial roots

Interpolation	
dsearch	Search Delaunay triangulation for nearest point
dsearchn	N-D nearest point search
griddata	Data gridding
griddata3	Data gridding and hypersurface fitting for 3-D data
griddatan	Data gridding and hypersurface fitting (dimension ≥ 2)
interp1	1-D data interpolation (table lookup)
interp1q	Quick 1-D linear interpolation
interp2	2-D data interpolation (table lookup)
interp3	3-D data interpolation (table lookup)
interpft	1-D interpolation using FFT method
interpn	N-D data interpolation (table lookup)
meshgrid	Generate X and Y arrays for 3-D plots
mkpp	Make piecewise polynomial
ndgrid	Generate arrays for N-D functions and interpolation
padecoef	Padé approximation of time delays
pchip	Piecewise Cubic Hermite Interpolating Polynomial (PCHIP)
ppval	Evaluate piecewise polynomial
spline	Cubic spline data interpolation
TriScatteredInterp	Interpolate scattered data
TriScatteredInterp	Interpolate scattered data
tsearch	Search for enclosing Delaunay triangle
tsearchn	N-D closest simplex search
unmkpp	Piecewise polynomial details

Discrete Math	
factor	Prime factors
factorial	Factorial function
gcd	Greatest common divisor
isprime	Array elements that are prime numbers
lcm	Least common multiple
nchoosek	Binomial coefficient or all combinations
perms	All possible permutations
primes	Generate list of prime numbers
rat, rats	Rational fraction approximation

The following tables provide only superficial information on additional categories of functions that are not individually listed here.

Nonlinear Numerical Methods	
Ordinary Differential Equations	Solve stiff and non-stiff differential equations, define the problem, set solver options, evaluate solution
Delay Differential Equations	Solve delay differential equations with constant and general delays, set solver options, evaluate solution
Boundary Value Problems	Solve boundary value problems for ordinary differential equations, set solver options, evaluate solution
Partial Differential Equations	Solve initial-boundary value problems for parabolic-elliptic PDEs, evaluate solution
Optimization	Find minimum of single and multivariable functions, solve nonnegative least-squares constraint problem

Sparse Matrices	
Elementary Sparse Matrices	Create random and nonrandom sparse matrices
Full to Sparse Conversion	Convert full matrix to sparse, sparse matrix to full
Sparse Matrix Manipulation	Test matrix for sparseness, get information on sparse matrix, allocate sparse matrix, apply function to nonzero elements, visualize sparsity pattern
Reordering Algorithms	Random, column, minimum degree, Dulmage-Mendelsohn, and reverse Cuthill-McKee permutations
Linear Algebra	Compute norms, eigenvalues, factorizations, least squares, structural rank
Linear Equations (Iterative Methods)	Methods for conjugate and biconjugate gradients, residuals, lower quartile
Tree Operations	Elimination trees, tree plotting, factorization analysis

Computational Geometry	
Delaunay Triangulation and Tessellation	Delaunay triangulation and tessellation, triangular surface and mesh plots
Convex Hull	Plot convex hull, plotting functions
Voronoi Diagrams	Plot Voronoi diagram, patch graphics object, plotting functions
Domain Generation	Generate arrays for 3-D plots, or for N-D functions and interpolation

A.11 GNU Octave

This section presents only a short summary about GNU Octave. Most of this material is based on information provided on the Octave Web page. Additional information about Octave is available on the Web page:

http://www.octave.org

GNU Octave is a high-level language, primarily intended for numerical computations. It provides a convenient command line interface for solving linear and nonlinear problems numerically, and for performing other numerical experiments using a language that is mostly compatible with MATLAB. It may also be used as a batch-oriented language.

GNU Octave is freely redistributable software. It may be redistributed and/or modified under the terms of the GNU General Public License (GPL) as published by the Free Software Foundation. Copyright ©1998-2010 John W. Eaton.

Octave was written by John W. Eaton and many others. Because Octave is free software users are encouraged to help make Octave more useful by writing and contributing additional functions for it, and by reporting any problems. GNU Octave is supported and developed by its user community.

On the left pane of the main page, click on the *download* link. Under *Binaries*, select the operating system being used (Linux, MacOS X, Windows, Sun Solaris. Click to download and select save to get the executable Octave installer.

Once the download is complete, run the executable file that was downloaded and saved. This will install Octave. When this is complete, Octave can be started. Figure A.6 shows the Octave command window. Since January 28, 2010, the stable release version is Octave-3.2.4.

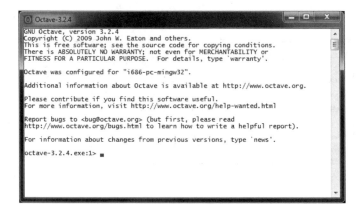

FIGURE A.6: The Octave window.

On the Octave window, to get help with individual commands and functions type:
```
help command
```
For example, after typing `help plot`, Octave responds with a new screen and information about the *plot* function. Figure A.7 illustrates how Octave displays the documentation about the *plot* function.

For a more detailed introduction to GNU Octave, please consult the manual. To read the manual at the prompt type `doc`. Figure A.8 shows the Octave

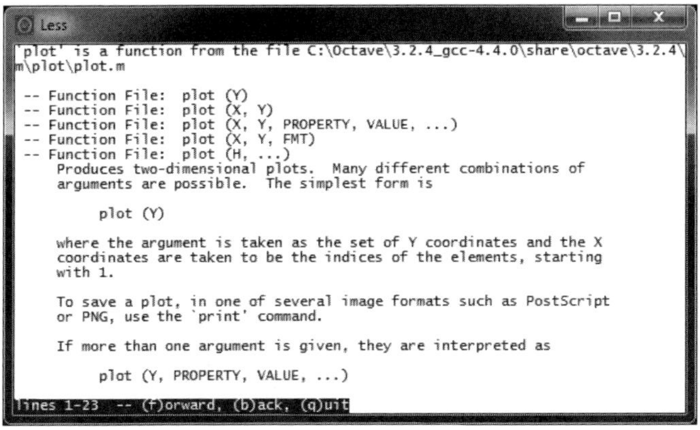

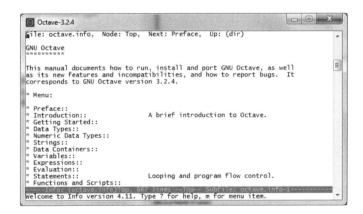

FIGURE A.7: Octave help documentation on *plot*.

window with the first page of the manual. By pressing the space bar, subsequent pages can be seen on the Octave window.

FIGURE A.8: Octave documentation manual.

Appendix B

Computer Systems

B.1 Introduction

This appendix presents an overview of computer systems. The basic concepts of hardware and software components of a computer system are briefly explained. The concepts of an operating system and application software are discussed as essential software components. This basic material is important in order to better understand the general concepts and principles of computational modeling.

B.2 Computer Systems

A *computer system* (also known as a computer) is basically an electronic machine that can perform various tasks and computations under control of the *software*. A computer system consists of two basic types of *components*:

- *Hardware components*, which are the electronic and electromechanical devices, such as the processors, memory modules, disk units, keyboard, screen, and other devices.

- *Software components*, such as the application programs, operating system, utilities, and others.

The software instructs the computer to execute the appropriate *instructions* and use input *data* to produce some desired results. The architecture of a computer system is the relationship among the various components. Computer systems have the following fundamental functions: processing, input, and output.

- *Input*, which enters data into the computer for processing. Data is entered into the computer by means of an input device such as the keyboard.

- *Processing*, which executes the instructions that have previously been stored in memory to perform some transformation and/or computation on the *data* carried out by one or more processors. Instructions and data must be placed first into memory in order for processing to proceed.
- *Output*, which transfers data from the computer to an output device such as the video screen or to a printer.

The data that has been entered into the computer is commonly known as *input data*. Similarly, the data that has been transferred out from the computer via an output device is known as *output data* or the resulting data.

B.2.1 Hardware Components

The basic structure of a computer normally consists of one or more of the following hardware components:

- The *central processing unit* (CPU), also called processor
- The *memory* also known as random access memory (RAM)
- The *massive storage* devices, which store large amounts of data and programs in permanent form
- The *input/output* (I/O) units
- The *system bus*, which provides interconnections for all the components of the system
- Additional devices

Figure B.1 illustrates this basic structure of a computer system. The computer shown has only one processor (CPU).

B.2.1.1 Processors

The processor (CPU) executes the *instructions* that are stored in memory. The processors perform arithmetic operations and data transfers. Larger computers have several processors. Most newer computer systems are designed and built with multicore chips.

The processor can only execute instructions if they are in a special form called *machine language*. This form of the instructions is dependent on the type of processor.

The speed of the processor is an important system parameter of the computer, and is usually measured in gigahertz (GHz). This is the number of machine cycles per second that the processor can handle. One or more machine cycles are required to execute a machine instruction.

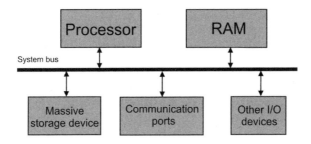

FIGURE B.1: Basic hardware structure of a computer system.

B.2.1.2 Main Memory

The *random access memory* (RAM) is a high-speed device for temporary storage of data and programs. The two relevant system parameters for memory are the total memory capacity and the memory access time.

Memory consists of a large number of storage cells. A very small amount of data or a small program instruction can be stored in a memory cell. Each of these cells has an associated memory location, which is known as a memory *address*. When the CPU needs to fetch some data from memory, it first gets the memory location of that data, and then it can access the data.

The *byte* is the smallest unit of storage. Every memory cell can store a byte of data or program. Memory capacity is measured by the total number of bytes it contains, measured in megabytes (MB) or gigabytes (GB).

Another important characteristic of main memory is the access time. The access time is the time interval that the CPU takes to fetch some data from a memory cell, or to store some data to a memory cell.

B.2.1.3 Storage Devices

The disks and tape units are external storage devices that are used to store programs and massive amounts of data, in a permanent form. The unit of storage for these devices is the byte, usually megabytes (MB) and gigabytes (GB). A typical hard disk can have a storage capacity of 260 GB or more. Other storage devices are compact discs (CD) and magnetic tape devices. The various disk and tape devices are connected to the disk and tape controllers (see Figure B.1). These storage devices are much slower than main memory; the CPU takes much more time to access data on a disk device than data on main memory.

The system parameters for these devices are similar to those for memory, total storage capacity and the access time.

B.2.1.4 Input Devices

Input devices, such as a *keyboard*, are used for entering data and/or programs into the computer. A storage device such as a disk can also be assigned as an input device.

B.2.1.5 Output Devices

Output devices, such as a video screen, are used for transferring data from inside the computer to the outside world.

The input and output devices are connected to the communication ports in the computer (see Figure B.1). For graphic applications, a video display device is connected to the graphic controller, which is a unit for connecting one or more video units and/or graphic devices.

The input and output devices are used extensively when the program maintains a dialog with the user while executing. This computer-user dialog is called user interaction.

B.2.1.6 Bus

All the hardware components are interconnected via electronic paths called the *bus*. Data and controlling signals are normally transferred to or from the CPU to the appropriate device through the bus. The speed capacity of the bus is another important system parameter that affects the overall performance of a computer system.

B.2.2 Computer Networks

A network consists of two or more computers interconnected in such a way that they can exchange data and programs. On a local area network (LAN), several small computers are connected to a larger computer called a server that stores the global files or databases and may include one or more shared printers. A local area network is limited to a relatively small geographical area, such as a building, a floor, or a university campus. Figure B.2 shows the simplified structure of a local area network. This example is a client-server system with one server.

A much larger type of network is known as a wide area network (WAN), it covers a large geographical region and connects local area networks located in

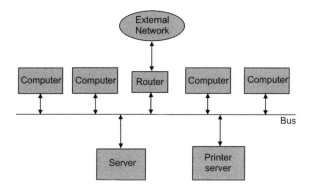

FIGURE B.2: Basic structure of a network.

various remote places. The Internet can be considered a public, international wide area network.

B.2.3 Software Components

The software consists of the collection of programs that execute in the computer. These programs perform computations, control, manage, and carry out other important tasks. There are two general types of software components:

- System software

- Application software

The *system software* is the set of programs that control the activities and functions of the various hardware components, programming tools and abstractions, and other utilities to monitor the state of the computer system. The system software forms an environment for the programmers to develop and execute their programs (collectively known as application software). There are actually two types of users: application programmers and end-users.

The most important part of system software is the operating system (OS). It directly controls and manages the hardware resources of the computer. Examples of operating systems are Unix, Windows, MacOS, OS/2, and others.

Application software consists of those programs that solve specific problems for the users. These programs execute under control of the system software. Application programs are developed by individuals and organizations for solving specific problems.

B.3 Operating Systems

An *operating system* is a large and complex set of programs that control the complete operation of a computer system and that provide a collection of services to other programs. Application and systems programmers directly or indirectly communicate with the operating system in order to request some of these services.

The purpose of an operating system involves two key goals:

1. Availability of a convenient, easy-to-use, and powerful set of services that are used by the users and the application programs in the computer system.

2. Management of the computer resources in the most efficient manner.

The services provided by the operating systems are implemented as a large set of functions that include scheduling of programs, memory management, device management, file management, network management, and other more advanced services related to protection and security. The operating system is also considered a huge resource manager.

The operating system must carry out these goals as efficiently as possible; otherwise, the performance of the system is affected.

The programmers and end-users communicate with the operating system and other system software. Users and programmers do not actually interact with the hardware.

B.3.1 Operating System User Interfaces

Users and application programmers can communicate with an operating system through its interfaces. There are three general levels of interfaces provided by an operating system, these are:

1. Graphical user interface (GUI), for example, the Windows desktop and the X-Window in Unix

2. Command level, also called the shell(s)

3. System-call interface, also called the application-programming interface (API)

The highest level is the graphical user interface because the user is presented with icons, menus and other graphical objects. With these objects, the user interacts with the operating system in a relatively easy and convenient manner, for example, using the click of a mouse.

The user at this level is completely separated from any intrinsic detail about the system. This level of operating system interface is not considered an essential part of the operating system, it is rather an add-on system software component.

The second level of interface, the shell, is a text-oriented interface. Advanced users and application programmers can communicate directly with the operating system.

The third level, the system-call interface (API), is needed by the application programs to request the various services provided by the operating system.

B.3.2 Contemporary Operating Systems

The two most widely used operating systems are Unix and Windows. These really represent two different families of operating systems that have evolved over the years. Several of the most widely used operating systems are:

- Windows (Microsoft Corporation): these include a family of systems: 98, CE, 2000, XP, Vista and Windows 7, and newer versions.
- Linux (Linus Torvalds, FSF GNU)
- Solaris (Sun Microsystems)
- IRIX (Silicon Graphics)
- OS2 (IBM)
- OS/390 (IBM)
- MacOS X (Apple)

B.3.2.1 Unix

Unix was originally introduced in 1974 by Ritchie and Thompson while working in AT&T Bell Labs. The operating system was developed on a small computer and two of the design goals were to be a small software system and to be easily portable. Many universities and research labs were users by 1980. Two versions became the best known systems: System V Unix (AT&T) and BSD Unix (University of California at Berkeley).

Unix became the dominant time-sharing OS and was used in small and large computers. It is one of the first operating systems written almost entirely in a high level programming language, C.

Linux belongs to the Unix family and was developed for personal computers. The basic part of the operating system was designed and implemented first by Linus Torvals in 1991. He released the source code on the Internet and invited designers and programmers to contribute their modifications and enhancements. Today Linux is freely available and has been implemented in many small and large computers. It has also become important in commercial software systems.

B.3.2.2 Microsoft Windows

The family of the Windows operating systems is considered one of the most widely used operating systems released by Microsoft. These systems were developed for personal computers that use the Intel-type microprocessors. This group of operating systems consists of Windows 95, 98, CE, XP, Vista, Windows 7 and others, which can be considered the smaller members of the family.

The larger Windows operating systems are designed for servers and include a comprehensive security model, more functionality facility for network support, improved and more powerful virtual memory management, and full implementation of the Windows API functions.

B.4 Summary

A computer system is an electronic machine with several devices connected to it. The computer system includes software and hardware components. An operating system is a large and complex software component of a computer system. It controls the major operations of user programs and manages all the hardware resources. This appendix has presented the description of the basic structure of a computer system and fundamental concepts of operating systems. Discussions on the general structure, user interfaces, and functions of an operating system have also been presented. Two examples of well-known operating systems are Unix and Microsoft Windows.

Key Terms		
computer system	hardware components	software components
input	processing	output
CPU	RAM	storage devices
I/O devices	bus	machine language
CPU speed	commands	byte
access time	networks	system software
application software	Linux	Unix
Windows	MacOS	OS interfaces
device manager	batch systems	interactive systems

Bibliography

[1] J. J. Banks, S. Carson, and B. Nelson. *Discrete-Event System Simulation*. Second ed. Englewood Cliffs, NJ: Prentice-Hall, 1996.

[2] John W. Eaton. *GNU Octave Manual*. Network Theory Limited, United Kingdom, 2002.

[3] Joyce Farrell. *Programming Logic and Design*. Second edition. Boston, MA: Course Technology Thomson Learning, 2002.

[4] Behrouz A. Forouzan. *Foundations of Computer Science: From Data Manipulation to Theory of Computation*. Pacific grove, CA: Brooks/Cole (Thomson), 2003.

[5] G. Fulford, P. Forrester, and A. Jones. *Modeling with Differential and Difference Equations*. New York: Cambridge University Press, 1997.

[6] Garrido, J. M. *Object-Oriented Discrete-Event Simulation with Java*. New York: Kluwer Academic/Plenum Publishers, 2001.

[7] Rod Haggarty. *Discrete Mathematics for Computing*. Addison Wesley (Pearson), Harlow, UK, 2002.

[8] Dan Kalman. *Elementary Mathematical Models*. Washington, DC: The Mathematical Association of America, 1997.

[9] Jack P. C. Kleijnen. *Simulation: A Statistical Perspective*. New York: Wiley, 1992.

[10] Roland E. Larsen, Robert P. Hostteller, and Bruce H. Edwards. *Brief Calculus with Applications*. Alternate third ed. Lexington, MA: D. C. Heath and Company, 1991.

[11] Averill M. Law and W. David Kelton. *Simulation Modeling and Analysis*. Third ed. New York: McGraw-Hill Higher Education, 2000.

[12] Charles F. Van Loan and K.-Y. Daisy Fan. *Insight Through Computing: A MATLAB Introduction to Computational Science and Engineering*. Philadelphia, PA: SIAM-Society for Industrial and Applied Mathematics, 2009.

[13] E. B. Magrab, S. Azarm, B. Balachandran, J. H. Duncan, K. E. Herold, and G. C. Walsh. *An Engineer's Guide to MATLAB: With Applications from Mechanical, Aerospace, Electrical, Civil, and Biological Systems Engineering*. Third ed. Upper Saddle River, NJ: Prentice Hall, Pearson, 2011.

[14] I Mitrani. *Simulation Techniques for Discrete Event Systems*. Cambridge: Cambridge University Press, 1982 (Reprinted 1986).

[15] Douglas Mooney and Randall Swift. *A Course in Mathematical Modeling*. Washington, DC: The Mathematical Association of America, 1999.

[16] Holly Moore. *MATLAB for Engineers*. Second ed. Upper Saddle River, NJ: Prentice Hall, Pearson, 2009.

[17] E. Part-Enander, A. Sjoberg, B. Melin, and P. Isaksson. *The MATLAB Handbook*. Harlow, UK: Addison-Wesley Longman, 1996.

[18] Harold J. Rood. *Logic and Structured Design for Computer Programmers*. Third edition. Pacific Grove, CA: Brooks/Cole (Thomson), 2001.

[19] Angela B. Shiflet and George W. Shiflet. *Introduction to Computational Science: Modeling and Simulation for the Sciences*. Princeton, NJ: Princeton University Press, 2006.

[20] David M. Smith. *Engineering Computation with MATLAB*. Boston, MA: Addison-Wesley, Pearson Education, 2010.

[21] Robert E. White. *Computational Mathematics: Models, Methods, and Analysis with MATLAB and MPI*. Boca raton, FL: Chapman and Hall/CRC. September 17, 2003.

[22] Wing, Jeannette M."Computational Thinking." *Communications of the ACM*. March 2006. Vol. 49, No. 3.

Index

abstract-world, 176
abstraction, 44, 85, 163, 185
accumulator, 120
algorithm, 83, 120
algorithm notation, 84
alternation, 99
analytical methods, 177
application software, 293
approximate values, 181
area, 67, 156
argument, 37, 171
arguments, 170
arithmetic, 93
array, 50, 131
array concatenation, 240, 245
array element, 136
array index, 135
array indexing, 247
array operation, 137
array size, 132
array slicing, 242
assignment, 92
assignment operator, 35
assumption, 179
assumptions, 175, 185
average, 182

boolean, 24
built-in functions, 37
byte, 291
bytecode, 26

calculations, 7
call, 168
case conversion, 255

case structure, 109, 111
character, 23
character array, 253
coefficients, 196
coefficients vector, 203, 206
collection, 50
colon operator, 51
column, 132
column vector, 239
command file, 126
command prompt, 33
commands, 48, 84, 167
comments, 38
compilation, 25
complex condition, 113
complex numbers, 104
complex plane, 104
complex roots, 207
components, 57
computational actions, 6
computational model, 21, 43, 83, 175
computational science, 44
computational thinking, 3, 44
computations, 21
computer implementation, 84
computer science, 44
computer system, 289
computer tools, 46
conceptual model, 61
condition, 101, 117, 120
conditional expressions, 101
conditions, 99
constant, 132
continuous data, 233
continuous model, 57

counter, 120, 124
CPU, 290
creating arrays, 132
cubic spline interpolation, 215
curve fitting, 211, 218, 221

data, 166, 289, 291
data decreases, 226
data description, 84
data increases, 226
data list, 50, 177
data points, 211
data structure, 239
data type, 22
data values, 22
decomposition, 64, 163
design structure, 99
design structures, 83, 87
deterministic model, 59
development process, 57
difference equation, 193
difference equations, 177
differences, 182, 187, 189
discrete data, 179
discrete instants, 58
discrete model, 58, 177
disks, 291
distance, 69
domain, 47
dot division, 242
dot multiplication, 241
dot powers, 242
double, 23
dynamic behavior, 63

electronic, 289
element, 52
elements, 131
environment, 57
estimate, 211
estimate data, 213
estimates, 211, 220
evaluate polynomial, 201

exponential functions, 233
exponential growth, 232
expression, 36
expression evaluation, 36
expressions, 93

factor, 225
factorial, 126
float, 23
flowcharts, 84, 85
for loop, 124
function, 47, 65, 150
function domain, 202
function invocation, 168
function range, 202
functional equation, 192, 232
functional equations, 177

geometric growth, 225
given data, 211
graphical methods, 176
graphical representation, 201
growth factor, 225, 226

hardware components, 289
heterogeneous collections, 259
high-level programming language, 24

identifier, 22
identifiers, 23, 91
if statement, 100
implementation, 60
independent variable, 48, 65, 180, 202
index, 131
input, 86, 93, 94, 289
input-output, 87
input/output, 99, 290
instructions, 21, 289, 290
integer, 23
interface, 166
intermediate, 211
intermediate data, 212
interpolation, 211
interpreter, 34

interval of x, 202
iterations, 120

keyboard, 292
keywords, 22

LAN, 292
languages, 91
least squares, 218
library function, 182, 190
library functions, 171
line slope, 65, 150
linear change, 189
linear interpolation, 211, 212
linear regression, 218
linear search, 145
linking, 25
local data, 166
local variables, 167
logarithm, 233
loop condition, 118
loop termination, 118
loops, 117
lower bound, 202

machine code, 25
maintenance, 30
massive storage, 290
mathematical expression, 192
mathematical methods, 176
mathematical model, 44, 60
mathematical representation, 47
matrices, 138
matrix, 131, 239
matrix operations, 246
memory, 290, 291
model, 21, 43, 106
model development, 59
modeling, 21, 43, 62
module interface, 164
modules, 64, 163
multidimensional array, 138

natural logarithm, 234

nonlinear interpolation, 215
nouns, 4
numerical methods, 176
numerical solution, 45

one-dimensional, 239
operands, 36
operating system, 289, 294
operating system goals, 294
operators, 36
order, 7
ordered list, 188
OS interfaces, 294
output, 86, 94, 290

parameters, 169
patterns, 16
percentage, 225
performance, 64
plotting, 249
polynomial, 201
polynomial degree, 201
polynomial expression, 218
position, 69
predefined constants, 35
problem description, 3
problem solution, 83
problem statement, 60
processing, 90, 290
processing symbol, 86
program, 21
programming language, 21, 24
programs, 293
prototypes, 30
pseudo-code, 84, 86

quadratic, 201
quadratic equation, 99, 196
quadratic growth, 187, 192
quadratic models, 193

RAM, 291
random variables, 59
range, 47

rate of change, 65, 66, 150
raw data, 211
real system, 62
real-world, 176
real-world problem, 175
regression, 211, 221
relational operators, 102
repeat-until, 90, 122
repetition, 87, 88, 99
repetition structure, 117
reserved words, 91
return, 171
return data, 170
reuse, 168
roots, 196
roots of function, 206
roots vector, 206
row, 132
row vector, 133, 239

scalar, 137
scientific notation, 37
screen, 292
script, 35
search condition, 145
secant, 66, 151
second differences, 190, 192
selection, 86–88, 99
semantic rules, 24
sequence, 87, 88, 99, 177, 189, 191
simulation, 61
simulation model, 57
simulation run, 64
slope of line, 66, 150
software, 21, 289
software components, 289
solve polynomial, 201
source program, 25
specification, 60, 175
spiral model, 30
spline interpolation, 211
start symbol, 85
statements, 91

stepwise refinement, 164
stochastic model, 59
stop symbol, 85
string, 23
string concatenation, 254
string conversion, 254
string matrix, 257
string vector, 253
strings, 253
subproblem, 163
substring, 254
summation, 68, 125, 157
syntax rules, 24
system, 43
system bus, 290
system software, 293

tangent, 66, 151
testing, 29
text characters, 253
time-dependent model, 177
top-down approach, 163
trace, 64
transformation, 86
trapezoid, 68, 157
two-dimensional, 239
two-dimensional arrays, 245

UML, 61
upper bound, 202

validation, 60
variable, 34
variable name, 35
vector, 50
vector addition, 241
vector assignment, 240
vector functions, 243
vector operations, 240
vector subtraction, 241
velocity, 69
verbs, 4
verification, 60

WAN, 292
waterfall model, 29
while loop, 90, 117, 118, 121
while statement, 118